Lecture Notes in Physics

Springer-Verlag
Berlin Heidelberg GmbH

The Editorial Policy for Proceedings

The series Lecture Notes in Physics reports new developments in physical research and teaching – quickly, informally, and at a high level. The proceedings to be considered for publication in this series should be limited to only a few areas of research, and these should be closely related to each other. The contributions should be of a high standard and should avoid lengthy redraftings of papers already published or about to be published elsewhere. As a whole, the proceedings should aim for a balanced presentation of the theme of the conference including a description of the techniques used and enough motivation for a broad readership. It should not be assumed that the published proceedings must reflect the conference in its entirety. (A listing or abstracts of papers presented at the meeting but not included in the proceedings could be added as an appendix.)

When applying for publication in the series Lecture Notes in Physics the volume's editor(s) should submit sufficient material to enable the series editors and their referees to make a fairly accurate evaluation (e.g. a complete list of speakers and titles of papers to be presented and abstracts). If, based on this information, the proceedings are (tentatively) accepted, the volume's editor(s), whose name(s) will appear on the title pages, should select the papers suitable for publication and have them refereed (as for a journal) when appropriate. As a rule discussions will not be accepted. The series editors and Springer-Verlag will normally not interfere with the detailed editing except in fairly obvious cases or on technical matters.

Final acceptance is expressed by the series editor in charge, in consultation with Springer-Verlag only after receiving the complete manuscript. It might help to send a copy of the authors' manuscripts in advance to the editor in charge to discuss possible revisions with him. As a general rule, the series editor will confirm his tentative acceptance if the final manuscript corresponds to the original concept discussed, if the quality of the contribution meets the requirements of the series, and if the final size of the manuscript does not greatly exceed the number of pages originally agreed upon. The manuscript should be forwarded to Springer-Verlag shortly after the meeting. In cases of extreme delay (more than six months after the conference) the series editors will check once more the timeliness of the papers. Therefore, the volume's editor(s) should establish strict deadlines, or collect the articles during the conference and have them revised on the spot. If a delay is unavoidable, one should encourage the authors to update their contributions if appropriate. The editors of proceedings are strongly advised to inform contributors about these points at an early stage.

The final manuscript should contain a table of contents and an informative introduction accessible also to readers not particularly familiar with the topic of the conference. The contributions should be in English. The volume's editor(s) should check the contributions for the correct use of language. At Springer-Verlag only the prefaces will be checked by a copy-editor for language and style. Grave linguistic or technical shortcomings may lead to the rejection of contributions by the series editors. A conference report should not exceed a total of 500 pages. Keeping the size within this bound should be achieved by a stricter selection of articles and not by imposing an upper limit to the length of the individual papers. Editors receive jointly 30 complimentary copies of their book. They are entitled to purchase further copies of their book at a reduced rate. As a rule no reprints of individual contributions can be supplied. No royalty is paid on Lecture Notes in Physics volumes. Commitment to publish is made by letter of interest rather than by signing a formal contract. Springer-Verlag secures the copyright for each volume.

The Production Process

The books are hardbound, and the publisher will select quality paper appropriate to the needs of the author(s). Publication time is about ten weeks. More than twenty years of experience guarantee authors the best possible service. To reach the goal of rapid publication at a low price the technique of photographic reproduction from a camera-ready manuscript was chosen. This process shifts the main responsibility for the technical quality considerably from the publisher to the authors. We therefore urge all authors and editors of proceedings to observe very carefully the essentials for the preparation of camera-ready manuscripts, which we will supply on request. This applies especially to the quality of figures and halftones submitted for publication. In addition, it might be useful to look at some of the volumes already published. As a special service, we offer free of charge LATEX and TEX macro packages to format the text according to Springer-Verlag's quality requirements. We strongly recommend that you make use of this offer, since the result will be a book of considerably improved technical quality. To avoid mistakes and time-consuming correspondence during the production period the conference editors should request special instructions from the publisher well before the beginning of the conference. Manuscripts not meeting the technical standard of the series will have to be returned for improvement.

For further information please contact Springer-Verlag, Physics Editorial Department II, Tiergartenstrasse 17, D-69121 Heidelberg, Germany

R. López-Peña   R. Capovilla   R. García-Pelayo
H. Waelbroeck   F. Zertuche (Eds.)

Complex Systems and Binary Networks

Guanajuato Lectures
Held at Guanajuato, México
16 – 22 January 1995

Springer

Editors

Ramón López-Peña
Henri Waelbroeck
Instituto de Ciencias Nucleares, UNAM
Apdo. Postal 70-543
México, D.F., 04510, México

Riccardo Capovilla
Depto. de Física, CINVESTAV IPN
Apdo. Postal 14-740
México, D.F., 07000, México

Ricardo García-Pelayo
Instituto de Física, UNAM
Apdo. Postal 20-364
México, D.F., 01000, México

Federico Zertuche
Instituto de Investigaciones en Matemáticas Aplicadas
y en Sistemas, Sección Cuernavaca, UNAM
Apdo. Postal 48-3
62251 Cuernavaca, Morelos, México

Die Deutsche Bibliothek - CIP-Einheitsaufnahme

Complex systems and binary networks : Guanajuato lectures,
held at Guanajuato, México, 16 - 22 January 1995 / R. López-
Peña ... (ed.).

(Lecture notes in physics ; 461)
ISBN 978-3-662-14057-4   ISBN 978-3-540-44937-9 (eBook)
DOI 10.1007/978-3-540-44937-9

NE: López-Peña, Ramón [Hrsg.]; GT

ISBN 978-3-662-14057-4

Typesetting: Camera-ready by the authors
SPIN: 10514988   55/3142-543210 - Printed on acid-free paper

Preface

A leading figure in Complex Systems once said: "Complex Systems are like beauty: You know it when you see it". This sentences gives rise, of course, to ample epistemological considerations. The most cheerful which comes to our mind is a remark due to Benoit Mandelbrot: At an early stage of the development of a subject, a certain vagueness in the definition is advantageous in that it doesn't limit future paths of evolution.

But we prefer to focus on the first sentence. Inspired by that sweeping definition, we have taken a direct, hands-on approach to this subject, which is wide open, multiple and open to joyful innovations. The origin of the list of topics is therefore our unbounded desire to learn and explore, which we hope the reader shares.

The mathematics of complex systems weaves its web over a wide range of fields. From the theory of information to knot theory, from neural computing to evolution theory, from galaxy formation to the birth of universes, most areas of modern physics stand to receive some new insight from complex systems theory.

Conversely, in order to understand complex systems one must draw inspiration from various sources and break free from overspecialization. That is the practical motivation for this work. To promote the cross-fertilization of ideas, we attempted to select a sampling of contributions such that each has something to learn from every other, and something to offer in return.

The book is structured so as to begin with those contributions which help to set up a general theoretical framework, and end with selected applications to particular areas of physics, in particular biophysics (RNA), statistical physics (Ising and Potts models) and astrophysics and cosmology.

In the attempts to measure complexity Chaitin's stands out for its naturalness. It is also deeply bound to Gödel's theorem, on the mathematics side, and to statistical mechanics and information theory on the physics side.

In the next section, Kauffman succeeds in giving a very clear introduction to knot theory, a subject which has a reputation for being intimidating, and in spelling out its applications to statistical mechanics and topological field theory.

To cap the theoretical structures and move into applications, Stadler offers a mathematical framework to accommodate all problems which can be formulated in terms of a "fitness landscape" over hypercubes, from the Travelling Salesman Problem to RNA secondary structures.

Derrida, a world authority on binary networks, gives us a solid introduction to this topic and to the statistical mechanics related to it, with an emphasis on the physical applications of complexity.

Finally, Smolin proposes a view of the Universe which challenges the imagination, from self-organizing models of galaxy formation to the birth of baby universes and the condensation of structure constants.

Of course no single book can possibly contain an exhaustive list of the developments of complex systems theory. Instead, we have attempted to provide a palette of colors, which the reader is invited to blend into a painting more faithful to his/her own interest and imagination.

Mexico City   R. García-Pelayo
August 1995   H. Waelbroeck

Contents

Part I
Randomness & Complexity in Pure Mathematics

Randomness & Complexity in Pure Mathematics

(Originally appeared in International Journal of Bifurcation and

Chaos 4 (1994), pp. 3–15)

G. J. Chaitin

IBM Research Division,
P.O. Box 704, Yorktown Heights, NY 10598, USA,
chaitin watson.ibm.com

Abstract. One normally thinks that everything that is true is true for a reason. I've found mathematical truths that are true for no reason at all. These mathematical truths are beyond the power of mathematical reasoning because they are accidental and random.

Using software written in *Mathematica* that runs on an IBM RS/6000 workstation, I constructed a perverse 200-page algebraic equation with a parameter N and 17,000 unknowns:

$$\text{Left-Hand-Side}(N) = \text{Right-Hand-Side}(N).$$

For each whole-number value of the parameter N, ask whether this equation has a finite or an infinite number of whole number solutions. The answers escape the power of mathematical reason because they are completely random and accidental.

This work is an extension of famous results of Gödel and Turing using ideas from a new field called algorithmic information theory.

1 Hilbert on the axiomatic method

Last month I was a speaker at a symposium on reductionism at Cambridge University where Turing did his work. I'd like to repeat the talk I gave there and explain how my work continues and extends Turing's. Two previous speakers had said bad things about David Hilbert. So I started by saying that in spite of what you might have heard in some of the previous lectures, Hilbert was not a twit!

Hilbert's idea is the culmination of two thousand years of mathematical tradition going back to Euclid's axiomatic treatment of geometry, going back to Leibniz's dream of a symbolic logic and Russell and Whitehead's monumental *Principia Mathematica.* Hilbert's dream was to once and for all clarify the methods of mathematical reasoning. Hilbert wanted to formulate a formal axiomatic system which would encompass all of mathematics.

Formal Axiomatic System
$\longrightarrow$
$\longrightarrow$
$\longrightarrow$

Hilbert emphasized a number of key properties that such a formal axiomatic system should have. It's like a computer programming language. It's a precise statement about the methods of reasoning, the postulates and the methods of inference that we accept as mathematicians. Furthermore Hilbert stipulated that the formal axiomatic system encompassing all of mathematics that he wanted to construct should be "consistent" and it should be "complete."

Formal Axiomatic System
$\longrightarrow$ consistent
$\longrightarrow$ complete
$\longrightarrow$

Consistent means that you shouldn't be able to prove an assertion and the contrary of the assertion.

Formal Axiomatic System
$\longrightarrow$ consistent $A \ \neg A$
$\longrightarrow$ complete
$\longrightarrow$

You shouldn't be able to prove A and not A. That would be very embarrassing.

Complete means that if you make a meaningful assertion you should be able to settle it one way or the other. It means that either A or not A should be a theorem, should be provable from the axioms using the rules of inference in the formal axiomatic system.

Formal Axiomatic System
$\longrightarrow$ consistent $A \ \neg A$
$\longrightarrow$ complete $A \ \neg A$
$\longrightarrow$

Consider a meaningful assertion A and its contrary not A. Exactly one of the two should be provable if the formal axiomatic system is consistent and complete.

A formal axiomatic system is like a programming language. There's an alphabet and rules of grammar, in other words, a formal syntax. It's a kind of thing that we are familiar with now. Look back at Russell and Whitehead's three enormous volumes full of symbols and you'll feel you're looking at a large computer program in some incomprehensible programming language.

Now there's a very surprising fact. Consistent and complete means only truth and all the truth. They seem like reasonable requirements. There's a funny consequence, though, having to do with something called the decision problem. In

German it's the Entscheidungsproblem.

Formal Axiomatic System
$\longrightarrow$ consistent $A \ \neg A$
$\longrightarrow$ complete $A \ \neg A$
$\longrightarrow$ decision problem

Hilbert ascribed a great deal of importance to the decision problem.

HILBERT
Formal Axiomatic System
$\longrightarrow$ consistent $A \ \neg A$
$\longrightarrow$ complete $A \ \neg A$
$\longrightarrow$ decision problem

Solving the decision problem for a formal axiomatic system is giving an algorithm that enables you to decide whether any given meaningful assertion is a theorem or not. A solution of the decision problem is called a decision procedure.

HILBERT
Formal Axiomatic System
$\longrightarrow$ consistent $A \ \neg A$
$\longrightarrow$ complete $A \ \neg A$
$\longrightarrow$ decision procedure

This sounds weird. The formal axiomatic system that Hilbert wanted to construct would have included all of mathematics: elementary arithmetic, calculus, algebra, everything. If there's a decision procedure, then mathematicians are out of work. This algorithm, this mechanical procedure, can check whether something is a theorem or not, can check whether it's true or not. So to require that there be a decision procedure for this formal axiomatic system sounds like you're asking for a lot.

However it's very easy to see that if it's consistent and it's complete that implies that there must be a decision procedure. Here's how you do it. You have a formal language with a finite alphabet and a grammar. And Hilbert emphasized that the whole point of a formal axiomatic system is that there must be a mechanical procedure for checking whether a purported proof is correct or not, whether it obeys the rules or not. That's the notion that mathematical truth should be objective so that everyone can agree whether a proof follows the rules or not.

So if that's the case you run through all possible proofs in size order, and look at all sequences of symbols from the alphabet one character long, two, three, four, a thousand, a thousand and one... a hundred thousand characters long. You apply the mechanical procedure which is the essence of the formal axiomatic system, to check whether each proof is valid. Most of the time, of course, it'll be nonsense, it'll be ungrammatical. But you'll eventually find every possible proof. It's like a million monkeys typing away. You'll find every possible proof, though only in principle of course. The number grows exponentially and this is something that you couldn't do in practice. You'd never get to proofs that are one page long.

But in principle you could run through all possible proofs, check which ones are valid, see what they prove, and that way you can systematically find all theorems. In other words, there is an algorithm, a mechanical procedure, for generating one by one every theorem that can be demonstrated in a formal axiomatic system. So if for every meaningful assertion within the system, either the assertion is a theorem or its contrary is a theorem, only one of them, then you get a decision procedure. To see whether an assertion is a theorem or not you just run through all possible proofs until you find the assertion coming out as a theorem or you prove the contrary assertion.

So it seems that Hilbert actually believed that he was going to solve once and for all, all mathematical problems. It sounds amazing, but apparently he did. He believed that he would be able to set down a consistent and complete formal axiomatic system for all of mathematics and from it obtain a decision procedure for all of mathematics. This is just following the formal, axiomatic tradition in mathematics.

But I'm sure he didn't think that it would be a practical decision procedure. The one I've outlined would only work in principle. It's exponentially slow, it's terribly slow! Totally impractical. But the idea was that if all mathematicians could agree whether a proof is correct and be consistent and complete, in principle that would give a decision procedure for automatically solving any mathematical problem. This was Hilbert's magnificent dream, and it was to be the culmination of Euclid and Leibniz, and Boole and Peano, and Russell and Whitehead.

Of course the only problem with this inspiring project is that it turned out to be impossible!

2 Gödel, Turing and Cantor's diagonal argument

Hilbert is indeed inspiring. His famous lecture in the year 1900 is a call to arms to mathematicians to solve a list of twenty-three difficult problems. As a young kid becoming a mathematician you read that list of twenty-three problems and Hilbert is saying that there is no limit to what mathematicians can do. We can solve a problem if we are clever enough and work at it long enough. He didn't believe that in principle there was any limit to what mathematics could achieve.

I think this is very inspiring. So did John von Neumann. When he was a young man he tried to carry through Hilbert's ambitious program. Because Hilbert couldn't quite get it all to work, in fact he started off just with elementary number theory, 1, 2, 3, 4, 5, ..., not even with real numbers at first.

And then in 1931 to everyone's great surprise (including von Neumann's), Gödel showed that it was impossible, that it couldn't be done, as I'm sure you all know.

Gödel 1931

This was the opposite of what everyone had expected. Von Neumann said it never occurred to him that Hilbert's program couldn't be carried out. Von Neumann admired Gödel enormously, and helped him to get a permanent position at the Institute for Advanced Study.

What Gödel showed was the following. Suppose that you have a formal axiomatic system dealing with elementary number theory, with 1, 2, 3, 4, 5 and addition and multiplication. And we'll assume that it's consistent, which is a minimum requirement—if you can prove false results it's really pretty bad. What Gödel showed was that if you assume that it's consistent, then you can show that it's incomplete. That was Gödel's result, and the proof is very clever and involves self-reference. Gödel was able to construct an assertion about the whole numbers that says of itself that it's unprovable. This was a tremendous shock. Gödel has to be admired for his intellectual imagination; everyone else thought that Hilbert was right.

However I think that Turing's 1936 approach is better.

Gödel 1931
Turing 1936

Gödel's 1931 proof is very ingenious, it's a real tour de force. I have to confess that when I was a kid trying to understand it, I could read it and follow it step by step but somehow I couldn't ever really feel that I was grasping it. Now Turing had a completely different approach.

Turing's approach I think it's fair to say is in some ways more fundamental. In fact, Turing did more than Gödel. Turing not only got as a corollary Gödel's result, he showed that there could be no decision procedure.

You see, if you assume that you have a formal axiomatic system for arithmetic and it's consistent, from Gödel you know that it can't be complete, but there still might be a decision procedure. There still might be a mechanical procedure which would enable you to decide if a given assertion is true or not. That was left open by Gödel, but Turing settled it. The fact that there cannot be a decision procedure is more fundamental and you get incompleteness as a corollary.

How did Turing do it? I want to tell you how he did it because that's the springboard for my own work. The way he did it, and I'm sure all of you have heard about it, has to do with something called the halting problem. In fact if you go back to Turing's 1936 paper you will not find the words "halting problem." But the idea is certainly there.

People also forget that Turing was talking about "computable numbers." The title of his paper is "On computable numbers, with an application to the Entscheidungsproblem." Everyone remembers that the halting problem is unsolvable and that comes from that paper, but not as many people remember that Turing was talking about computable real numbers. My work deals with computable and dramatically uncomputable real numbers. So I'd like to refresh your memory how Turing's argument goes.

Turing's argument is really what destroys Hilbert's dream, and it's a simple argument. It's just Cantor's diagonal procedure (for those of you who know what that is) applied to the computable real numbers. That's it, that's the whole idea in a nutshell, and it's enough to show that Hilbert's dream, the culmination of two thousand years of what mathematicians thought mathematics was about, is wrong. So Turing's work is tremendously deep.

What is Turing's argument? A real number, you know $3.1415926\cdots$, is a length measured with arbitrary precision, with an infinite number of digits. And a computable real number said Turing is one for which there is a computer program or algorithm for calculating the digits one by one. For example, there are programs for π, and there are algorithms for solutions of algebraic equations with integer coefficients. In fact most of the numbers that you actually find in analysis are computable. However they're the exception, if you know set theory, because the computable reals are denumerable and the reals are nondenumerable (you don't have to know what that means). That's the essence of Turing's idea.

The idea is this. You list all possible computer programs. At that time there were no computer programs, and Turing had to invent the Turing machine, which was a tremendous step forward. But now you just say, imagine writing a list with every possible computer program.

p_1 **Gödel 1931**
p_2 **Turing 1936**
p_3
p_4
p_5
p_6
$\vdots$

If you consider computer programs to be in binary, then it's natural to think of a computer program as a natural number. And next to each computer program, the first one, the second one, the third one, write out the real number that it computes if it computes a real (it may not). But if it prints out an infinite number of digits, write them out. So maybe it's 3.1415926 and here you have another and another and another:

p_1 $3.1415926\cdots$ **Gödel 1931**
p_2 $\cdots$ **Turing 1936**
p_3 $\cdots$
p_4 $\cdots$
p_5 $\cdots$
p_6 $\cdots$
$\vdots$

So you make this list. Maybe some of these programs don't print out an infinite number of digits, because they're programs that halt or that have an error in them and explode. But then there'll just be a blank line in the list.

$$p_1 \quad 3.1415926\cdots \quad \textbf{Gödel } 1931$$
$$p_2 \quad \cdots \quad\quad\quad\quad \textbf{Turing } 1936$$
$$p_3 \quad \cdots$$
$$p_4 \quad \cdots$$
$$p_5$$
$$p_6 \quad \cdots$$
$$\vdots$$

It's not really important—let's forget about this possibility.

Following Cantor, Turing says go down the diagonal and look at the first digit of the first number, the second digit of the second, the third...

$$p_1 \quad -.\underline{d_{11}}d_{12}d_{13}d_{14}d_{15}d_{16}\cdots \quad \textbf{Gödel } 1931$$
$$p_2 \quad -.d_{21}\underline{d_{22}}d_{23}d_{24}d_{25}d_{26}\cdots \quad \textbf{Turing } 1936$$
$$p_3 \quad -.d_{31}d_{32}\underline{d_{33}}d_{34}d_{35}d_{36}\cdots$$
$$p_4 \quad -.d_{41}d_{42}d_{43}\underline{d_{44}}d_{45}d_{46}\cdots$$
$$p_5$$
$$p_6 \quad -.d_{61}d_{62}d_{63}d_{64}d_{65}\underline{d_{66}}\cdots$$
$$\vdots$$

Well actually it's the digits after the decimal point. So it's the first digit after the decimal point of the the first number, the second digit after the decimal point of the second, the third digit of the third number, the fourth digit of the fourth, the fifth digit of the fifth. And it doesn't matter if the fifth program doesn't put out a fifth digit, it really doesn't matter.

What you do is you change these digits. Make them different. Change every digit on the diagonal. Put these changed digits together into a new number with a decimal point in front, a new real number. That's Cantor's diagonal procedure. So you have a digit which you choose to be different from the first digit of the first number, the second digit of the second, the third of the third, and you put these together into one number.

$$p_1 \quad -.\underline{d_{11}}d_{12}d_{13}d_{14}d_{15}d_{16}\cdots \quad\quad\quad \textbf{Gödel } 1931$$
$$p_2 \quad -.d_{21}\underline{d_{22}}d_{23}d_{24}d_{25}d_{26}\cdots \quad\quad\quad \textbf{Turing } 1936$$
$$p_3 \quad -.d_{31}d_{32}\underline{d_{33}}d_{34}d_{35}d_{36}\cdots$$
$$p_4 \quad -.d_{41}d_{42}d_{43}\underline{d_{44}}d_{45}d_{46}\cdots$$
$$p_5$$
$$p_6 \quad -.d_{61}d_{62}d_{63}d_{64}d_{65}\underline{d_{66}}\cdots$$
$$\vdots$$
$$.\neq d_{11}\neq d_{22}\neq d_{33}\neq d_{44}\neq d_{55}\neq d_{66}\cdots$$

This new number cannot be in the list because of the way it was constructed. Therefore it's an uncomputable real number. How does Turing go on from here to the halting problem? Well, just ask yourself **why** can't you compute it? I've explained how to get this number and it looks like you could almost do it. To compute the Nth digit of this number, you get the Nth computer program (you can certainly do that) and then you start it running until it puts out an Nth digit, and at that point you change it. Well what's the problem? That sounds easy.

The problem is, what happens if the Nth computer program never puts out an Nth digit, and you sit there waiting? And that's the halting problem—you cannot decide whether the Nth computer program will ever put out an Nth digit! This is how Turing got the unsolvability of the halting problem. Because if you could solve the halting problem, then you could decide if the Nth computer program ever puts out an Nth digit. And if you could do that then you could actually carry out Cantor's diagonal procedure and compute a real number which has to differ from any computable real. That's Turing's original argument.

Why does this explode Hilbert's dream? What has Turing proved? That there is no algorithm, no mechanical procedure, which will decide if the Nth computer program ever outputs an Nth digit. Thus there can be no algorithm which will decide if a computer program is a special case). program ever halts (finding the Nth digit put out by the Nth program is a special case). Well, what Hilbert wanted was a formal axiomatic system from which all mathematical truth should follow, only mathematical truth, and all mathematical truth. If Hilbert could do that, it would give us a mechanical procedure to decide if a computer program will ever halt. Why?

You just run through all possible proofs until you either find a proof that the program halts or you find a proof that it never halts. So if Hilbert's dream of a finite set of axioms from which all of mathematical truth should follow were possible, then by running through all possible proofs checking which ones are correct, you would be able to decide if any computer program halts. In principle you could. But you **can't** by Turing's very simple argument which is just Cantor's diagonal argument applied to the computable reals. That's how simple it is!

Gödel's proof is ingenious and difficult. Turing's argument is so fundamental, so deep, that everything seems natural and inevitable. But of course he's building on Gödel's work.

3. The halting probability and algorithmic randomness

The reason I talked to you about Turing and computable reals is that I'm going to use a different procedure to construct an uncomputable real, a much more uncomputable real than Turing does.

$$
\begin{array}{lll}
p_1 & -.\underline{d_{11}}d_{12}d_{13}d_{14}d_{15}d_{16}\cdots & \textbf{Gödel 1931} \\
p_2 & -.d_{21}\underline{d_{22}}d_{23}d_{24}d_{25}d_{26}\cdots & \textbf{Turing 1936} \\
p_3 & -.d_{31}d_{32}\underline{d_{33}}d_{34}d_{35}d_{36}\cdots & \text{uncomputable reals} \\
p_4 & -.d_{41}d_{42}d_{43}\underline{d_{44}}d_{45}d_{46}\cdots & \\
p_5 & & \\
p_6 & -.d_{61}d_{62}d_{63}d_{64}d_{65}\underline{d_{66}}\cdots & \\
\vdots & &
\end{array}
$$

$$. \neq d_{11} \neq d_{22} \neq d_{33} \neq d_{44} \neq d_{55} \neq d_{66} \cdots$$

And that's how we're going to get into much worse trouble.

How do I get a much more uncomputable real? (And I'll have to tell you how uncomputable it is.) Well, not with Cantor's diagonal argument. I get this number, which I like to call Ω, like this:

$$\Omega = \sum_{p \text{ halts}} 2^{-|p|}$$

This is just the halting probability. It's sort of a mathematical pun. Turing's fundamental result is that the halting problem is unsolvable—there is no algorithm that'll settle the halting problem. My fundamental result is that the halting probability is algorithmically irreducible or algorithmically random.

What exactly is the halting probability? I've written down an expression for it:

$$\Omega = \sum_{p \text{ halts}} 2^{-|p|}$$

Instead of looking at individual programs and asking whether they halt, you put all computer programs together in a bag. If you generate a computer program at random by tossing a coin for each bit of the program, what is the chance that the program will halt? You're thinking of programs as bit strings, and you generate each bit by an independent toss of a fair coin, so if a program is N bits long, then the probability that you get that particular program is 2^{-N}. Any program p that halts contributes $2^{-|p|}$, two to the minus its size in bits, the number of bits in it, to this halting probability.

By the way there's a technical detail which is very important and didn't work in the early version of algorithmic information theory. You couldn't write this:

$$\Omega = \sum_{p \text{ halts}} 2^{-|p|}$$

It would give infinity. The technical detail is that no extension of a valid program is a valid program. Then this sum

$$\sum_{p \text{ halts}} 2^{-|p|}$$

turns out to be between zero and one. Otherwise it turns out to be infinity. It only took ten years until I got it right. The original 1960s version of algorithmic information theory is wrong. One of the reasons it's wrong is that you can't even define this number

$$\Omega = \sum_{p \text{ halts}} 2^{-|p|}$$

In 1974 I redid algorithmic information theory with "self-delimiting" programs and then I discovered the halting probability Ω.

Okay, so this is a probability between zero and one

$$0 < \Omega = \sum_{p \text{ halts}} 2^{-|p|} < 1$$

like all probabilities. The idea is you generate each bit of a program by tossing a coin and ask what is the probability that it halts. This number Ω, this halting probability, is not only an uncomputable real—Turing already knew how to do that. It is uncomputable in the worst possible way. Let me give you some clues how uncomputable it is.

Well, one thing is it's algorithmically incompressible. If you want to get the first N bits of Ω out of a computer program, if you want a computer program that will print out the first N bits of Ω and then halt, that computer program has to be N bits long. Essentially you're only printing out constants that are in the program. You cannot squeeze the first N bits of Ω. This

$$0 < \Omega = \sum_{p \text{ halts}} 2^{-|p|} < 1$$

is a real number, you could write it in binary. And if you want to get out the first N bits from a computer program, essentially you just have to put them in. The program has to be N bits long. That's irreducible algorithmic information. There is no concise description.

Now that's an abstract way of saying things. Let me give a more concrete example of how random Ω is. Émile Borel at the turn of this century was one of the founders of probability theory.

$$0 < \Omega = \sum_{p \text{ halts}} 2^{-|p|} < 1$$

Émile Borel

QUESTION: Can I ask a very simple question before you get ahead?
ANSWER: Sure.
QUESTION: I can't see why Ω should be a probability. What if the two one-bit programs both halt? I mean, what if the two one-bit programs both halt and then some other program halts. Then Ω is greater than one and not a probability.
ANSWER: I told you no extension of a valid program is a valid program.

QUESTION: Oh right, no other programs can halt.

ANSWER: The two one-bit programs would be all the programs there are. That's the reason this number

$$0 < \Omega = \sum_{p \text{ halts}} 2^{-|p|} < 1$$

can't be defined if you think of programs in the normal way.

So here we have Émile Borel, and he talked about something he called a normal number.

$$0 < \Omega = \sum_{p \text{ halts}} 2^{-|p|} < 1$$

Émile Borel — normal reals

What is a normal real number? People have calculated π out to a billion digits, maybe two billion. One of the reasons for doing this, besides that it's like climbing a mountain and having the world record, is the question of whether each digit occurs the same number of times. It looks like the digits 0 through 9 each occur 10% of the time in the decimal expansion of π. It looks that way, but nobody can prove it. I think the same is true for $\sqrt{2}$, although that's not as popular a number to ask this about.

Let me describe some work Borel did around the turn of the century when he was pioneering modern probability theory. Pick a real number in the unit interval, a real number with a decimal point in front, with no integer part. If you pick a real number in the unit interval, Borel showed that with probability one it's going to be "normal." Normal means that when you write it in decimal each digit will occur in the limit exactly 10% of the time, and this will also happen in any other base. For example in binary 0 and 1 will each occur in the limit exactly 50% of the time. Similarly with blocks of digits. This was called an absolutely normal real number by Borel, and he showed that with probability one if you pick a real number at random between zero and one it's going to have this property. There's only one problem. He didn't know whether π is normal, he didn't know whether $\sqrt{2}$ is normal. In fact, he couldn't exhibit a single individual example of a normal real number.

The first example of a normal real number was discovered by a friend of Alan Turing's at Cambridge called David Champernowne, who is still alive and who's a well-known economist. Turing was impressed with him—I think he called him "Champ"—because Champ had published this in a paper as an undergraduate. This number is known as Champernowne's number. Let me show you Champernowne's number.

$$0 < \Omega = \sum_{p \text{ halts}} 2^{-|p|} < 1$$

Émile Borel — normal reals
Champernowne
.01234567891011121314⋯99100101⋯

It goes like this. You write down a decimal point, then you write 0, 1, 2, 3, 4, 5, 6, 7, 8, 9, then 10, 11, 12, 13, 14 until 99, then 100, 101. And you keep going in this funny way. This is called Champernowne's number and Champernowne showed that it's normal in base ten, only in base ten. Nobody knows if it's normal in other bases, I think it's still open. In base ten though, not only will the digits 0 through 9 occur exactly 10% of the time in the limit, but each possible block of two digits will occur exactly 1% of the time in the limit, each block of three digits will occur exactly .1% of the time in the limit, etc. That's called being normal in base ten. But nobody knows what happens in other bases.

The reason I'm saying all this is because it follows from the fact that the halting probability Ω is algorithmically irreducible information that this

$$0 < \Omega = \sum_{p \text{ halts}} 2^{-|p|} < 1$$

·is normal in any base. That's easy to prove using ideas about coding and compressing information that go back to Shannon. So here we finally have an example of an absolutely normal number. I don't know how natural you think it is, but it is a specific real number that comes up and is normal in the most demanding sense that Borel could think of. Champernowne's number couldn't quite do that.

This number Ω is in fact random in many more senses. I would say it this way. It cannot be distinguished from the result of independent tosses of a fair coin. In fact this number

$$0 < \Omega = \sum_{p \text{ halts}} 2^{-|p|} < 1$$

shows that you have total randomness and chaos and unpredictability and lack of structure in pure mathematics! The same way that all it took for Turing to destroy Hilbert's dream was the diagonal argument, you just write down this expression

$$0 < \Omega = \sum_{p \text{ halts}} 2^{-|p|} < 1$$

and this shows that there are regions of pure mathematics where reasoning is totally useless, where you're up against an impenetrable wall. This is all it takes. It's just this halting probability.

Why do I say this? Well, let's say you want to use axioms to prove what the bits of this number Ω are. I've already told you that it's uncomputable—right?— like the number that Turing constructs using Cantor's diagonal argument. So we know there is no algorithm which will compute digit by digit or bit by bit this number Ω. But let's try to prove what individual bits are using a formal axiomatic system. What happens?

The situation is very, very bad. It's like this. Suppose you have a formal axiomatic system which is N bits of formal axiomatic system (I'll explain what this means more precisely later). It turns out that with a formal axiomatic system of complexity N, that is, N bits in size, you can prove what the positions and values are of at most $N + c$ bits of Ω.

Now what do I mean by formal axiomatic system N bits in size? Well, remember that the essence of a formal axiomatic system is a mechanical procedure for checking whether a formal proof follows the rules or not. It's a computer program. Of course in Hilbert's days there were no computer programs, but after Turing invented Turing machines you could finally specify the notion of computer program exactly, and of course now we're very familiar with it.

So the proof-checking algorithm which is the essence of any formal axiomatic system in Hilbert's sense is a computer program, and just see how many bits long this computer program is.[1] That's essentially how many bits it takes to specify the rules of the game, the axioms and postulates and the rules of inference. If that's N bits, then you may be able to prove say that the first bit of Ω in binary is 0, that the second bit is 1, that the third bit is 0, and then there might be a gap, and you might be able to prove that the thousandth bit is 1. But you're only going to be able to settle N cases if your formal axiomatic system is an N-bit formal axiomatic system.

Let me try to explain better what this means. It means that you can only get out as much as you put in. If you want to prove whether an individual bit in a specific place in the binary expansion of the real number Ω is a 0 or a 1, essentially the only way to prove that is to take it as a hypothesis, as an axiom, as a postulate. It's irreducible mathematical information. That's the key phrase that really gives the whole idea.

Irreducible Mathematical Information

$$0 < \Omega = \sum_{p \text{ halts}} 2^{-|p|} < 1$$

Émile Borel — normal reals
Champernowne
.01234567891011121314$\cdots$99100101$\cdots$

Okay, so what have we got? We have a rather simple mathematical object that completely escapes us. Ω's bits have no structure. There is no pattern, there is no structure that we as mathematicians can comprehend. If you're interested in proving what individual bits of this number at specific places are, whether they're 0 or 1, reasoning is completely useless. Here mathematical reasoning is irrelevant and can get nowhere. As I said before, the only way a formal axiomatic system can get out these results is essentially just to put them in as assumptions, which means you're not using reasoning. After all, anything can be demonstrated by taking it as a postulate that you add to your set of axioms. So this is a worst possible case—this is irreducible mathematical information. Here is a case where there is no structure, there are no correlations, there is no pattern that we can perceive.

---

[1] Technical Note: Actually, it's best to think of the complexity of a formal axiomatic system as the size in bits of the computer program that enumerates the set of all theorems.

4. Randomness in arithmetic

Okay, what does this have to do with randomness in arithmetic? Now we're going back to Gödel—I skipped over him rather quickly, and now let's go back.

Turing says that you cannot use proofs to decide whether a program will halt. You can't always prove that a program will halt or not. That's how he destroys Hilbert's dream of a universal mathematics. I get us into more trouble by looking at a different kind of question, namely, can you prove that the fifth bit of this particular real number

$$0 < \Omega = \sum_{p \text{ halts}} 2^{-|p|} < 1$$

is a 0 or a 1, or that the eighth bit is a 0 or a 1. But these are strange-looking questions. Who had ever heard of the halting problem in 1936? These are not the kind of things that mathematicians normally worry about. We're getting into trouble, but with questions rather far removed from normal mathematics.

Even though you can't have a formal axiomatic system which can always prove whether a program halts or not, it might be good for everything else and then you could have an **amended** version of Hilbert's dream. And the same with the halting probability Ω. If the halting problem looks a little bizarre, and it certainly did in 1936, well, Ω is brand new and certainly looks bizarre. Who ever heard of a halting probability? It's not the kind of thing that mathematicians normally do. So what do I care about all these incompleteness results!

Well, Gödel had already faced this problem with his assertion which is true but unprovable. It's an assertion which says of itself that it's unprovable. That kind of thing also never comes up in real mathematics. One of the key elements in Gödel's proof is that he managed to construct an **arithmetical** assertion which says of itself that it's unprovable. It was getting this self-referential assertion to be in elementary number theory which took so much cleverness.

There's been a lot of work building on Gödel's work, showing that problems involving computations are equivalent to arithmetical problems involving whole numbers. A number of names come to mind. Julia Robinson, Hilary Putnam and Martin Davis did some of the important work, and then a key result was found in 1970 by Yuri Matijasevič. He constructed a diophantine equation, which is an algebraic equation involving only whole numbers, with a lot of variables. One of the variables, K, is distinguished as a parameter. It's a polynomial equation with integer coefficients and all of the unknowns have to be whole numbers— that's a diophantine equation. As I said, one of the unknowns is a parameter. Matijasevič's equation has a solution for a particular value of the parameter K if and only if the Kth computer program halts.

In the year 1900 Hilbert had asked for an algorithm which will decide whether a diophantine equation, an algebraic equation involving only whole numbers, has a solution. This was Hilbert's tenth problem. It was tenth is his famous list of twenty-three problems. What Matijasevič showed in 1970 was that this is equivalent to deciding whether an arbitrary computer program halts. So Turing's

halting problem is exactly as hard as Hilbert's tenth problem. It's exactly as
hard to decide whether an arbitrary program will halt as to decide whether an
arbitrary algebraic equation in whole numbers has a solution. Therefore there is
no algorithm for doing that and Hilbert's tenth problem cannot be solved—that
was Matijasevič's 1970 result.

Matijasevič has gone on working in this area. In particular there is a piece
of work he did in collaboration with James Jones in 1984. I can use it to follow in Gödel's footsteps, to follow Gödel's example. You see, I've shown that
there's complete randomness, no pattern, lack of structure, and that reasoning
is completely useless, if you're interested in the individual bits of this number

$$0 < \Omega = \sum_{p \text{ halts}} 2^{-|p|} < 1$$

Following Gödel, let's convert this into something in elementary number theory. Because if you can get into all this trouble in elementary number theory,
that's the bedrock. Elementary number theory, 1, 2, 3, 4, 5, addition and multiplication, that goes back to the ancient Greeks and it's the most solid part of all
of mathematics. In set theory you're dealing with strange objects like large cardinals, but here you're not even dealing with derivatives or integrals or measure,
only with whole numbers. And using the 1984 results of Jones and Matijasevič I
can indeed dress up Ω arithmetically and get randomness in elementary number
theory.[2]

What I get is an exponential diophantine equation with a parameter. "Exponential diophantine equation" just means that you allow variables in the exponents. In contrast, what Matijasevič used to show that Hilbert's tenth problem
is unsolvable is just a polynomial diophantine equation, which means that the
exponents are always natural number constants. I have to allow X^Y. It's not
known yet whether I actually need to do this. It might be the case that I can
manage with a polynomial diophantine equation. It's an open question, I believe
that it's not settled yet. But for now, what I have is an exponential diophantine
equation with seventeen thousand variables. This equation is two-hundred pages
long and again one variable is the parameter.

This is an equation where every constant is a whole number, a natural number, and all the variables are also natural numbers, that is, positive integers.
(Actually **non-negative** integers.) One of the variables is a parameter, and you
change the value of this parameter—take it to be 1, 2, 3, 4, 5. Then you ask,
does the equation have a finite or infinite number of solutions? My equation is
constructed so that it has a finite number of solutions if a particular individual
bit of Ω is a 0, and it has an infinite number of solutions if that bit is a 1. So
deciding whether my exponential diophantine equation in each individual case
has a finite or infinite number of solutions is exactly the same as determining

---

[2] To obtain the *Mathematica* software for doing this, send e-mail to chaitin @
watson.ibm.com.

what an individual bit of this

$$0 < \Omega = \sum_{p \text{ halts}} 2^{-|p|} < 1$$

halting probability is. And this is completely intractable because Ω is irreducible mathematical information.

Let me emphasize the difference between this and Matijasevič's work on Hilbert's tenth problem. Matijasevič showed that there is a polynomial diophantine equation with a parameter with the following property: You vary the parameter and ask, does the equation have a solution? That turns out to be equivalent to Turing's halting problem, and therefore escapes the power of mathematical reasoning, of formal axiomatic reasoning.

How does this differ from what I do? I use an exponential diophantine equation, which means I allow variables in the exponent. Matijasevič only allows constant exponents. The big difference is that Hilbert asked for an algorithm to decide if a diophantine equation has a solution. The question I have to ask to get randomness in elementary number theory, in the arithmetic of the natural numbers, is slightly more sophisticated. Instead of asking whether there is a solution, I ask whether there are a finite or infinite number of solutions—a more abstract question. This difference is necessary.

My two-hundred page equation is constructed so that it has a finite or infinite number of solutions depending on whether a particular bit of the halting probability is a 0 or a 1. As you vary the parameter, you get each individual bit of Ω. Matijasevič's equation is constructed so that it has a solution if and only if a particular program ever halts. As you vary the parameter, you get each individual computer program.

Thus even in arithmetic you can find Ω's absolute lack of structure, Ω's randomness and irreducible mathematical information. Reasoning is completely powerless in those areas of arithmetic. My equation shows that this is so. As I said before, to get this equation I use ideas that start in Gödel's original 1931 paper. But it was Jones and Matijasevič's 1984 paper that finally gave me the tool that I needed.

So that's why I say that there is randomness in elementary number theory, in the arithmetic of the natural numbers. This is an impenetrable stone wall, it's a worst case. ¿From Gödel we knew that we couldn't get a formal axiomatic system to be complete. We knew we were in trouble, and Turing showed us how basic it was, but Ω is an extreme case where reasoning fails completely.

I won't go into the details, but let me talk in vague information-theoretic terms. Matijasevič's equation gives you N arithmetical questions with yes/no answers which turn out to be only $\log N$ bits of algorithmic information. My equation gives you N arithmetical questions with yes/no answers which are irreducible, incompressible mathematical information.

5. Experimental mathematics

Okay, let me say a little bit in the minutes I have left about what this all means.

First of all, the connection with physics. There was a big controversy when quantum mechanics was developed, because quantum theory is nondeterministic. Einstein didn't like that. He said, "God doesn't play dice!" But as I'm sure you all know, with chaos and nonlinear dynamics we've now realized that even in classical physics we get randomness and unpredictability. My work is in the same spirit. It shows that pure mathematics, in fact even elementary number theory, the arithmetic of the natural numbers, 1, 2, 3, 4, 5, is in the same boat. We get randomness there too. So, as a newspaper headline would put it, God not only plays dice in quantum mechanics and in classical physics, but even in pure mathematics, even in elementary number theory. So if a new paradigm is emerging, randomness is at the heart of it. By the way, randomness is also at the heart of quantum field theory, as virtual particles and Feynman path integrals (sums over all histories) show very clearly. So my work fits in with a lot of work in physics, which is why I often get invited to talk at physics meetings.

However the really important question isn't physics, it's mathematics. I've heard that Gödel wrote a letter to his mother who stayed in Europe. You know, Gödel and Einstein were friends at the Institute for Advanced Study. You'd see them walking down the street together. Apparently Gödel wrote a letter to his mother saying that even though Einstein's work on physics had really had a tremendous impact on how people did physics, he was disappointed that his work had not had the same effect on mathematicians. It hadn't made a difference in how mathematicians actually carried on their everyday work. So I think that's the key question: How should you really do mathematics?

I'm claiming I have a much stronger incompleteness result. If so maybe it'll be clearer whether mathematics should be done the ordinary way. What is the ordinary way of doing mathematics? In spite of the fact that everyone knows that any finite set of axioms is incomplete, how do mathematicians actually work? Well suppose you have a conjecture that you've been thinking about for a few weeks, and you believe it because you've tested a large number of cases on a computer. Maybe it's a conjecture about the primes and for two weeks you've tried to prove it. At the end of two weeks you don't say, well obviously the reason I haven't been able to show this is because of Gödel's incompleteness theorem! Let us therefore add it as a new axiom! But if you took Gödel's incompleteness theorem very seriously this might in fact be the way to proceed. Mathematicians will laugh but physicists actually behave this way.

Look at the history of physics. You start with Newtonian physics. You cannot get Maxwell's equations from Newtonian physics. It's a new domain of experience—you need new postulates to deal with it. As for special relativity, well, special relativity is almost in Maxwell's equations. But Schrödinger's equation does not come from Newtonian physics and Maxwell's equations. It's a new domain of experience and again you need new axioms. So physicists are used to the idea that when you start experimenting at a smaller scale, or with new phenomena, you may need new principles to understand and explain what's going on.

Now in spite of incompleteness mathematicians don't behave at all like physicists do. At a subconscious level they still assume that the small number of principles, of postulates and methods of inference, that they learned early as mathematics students, are enough. In their hearts they believe that if you can't prove a result it's your own fault. That's probably a good attitude to take rather than to blame someone else, but let's look at a question like the Riemann hypothesis. A physicist would say that there is ample experimental evidence for the Riemann hypothesis and would go ahead and take it as a working assumption.

What is the Riemann hypothesis? There are many unsolved questions involving the distribution of the prime numbers that can be settled if you assume the Riemann hypothesis. Using computers people check these conjectures and they work beautifully. They're neat formulas but nobody can prove them. A lot of them follow from the Riemann hypothesis. To a physicist this would be enough: It's useful, it explains a lot of data. Of course a physicist then has to be prepared to say "Oh oh, I goofed!" because an experiment can subsequently contradict a theory. This happens very often.

In particle physics you throw up theories all the time and most of them quickly die. But mathematicians don't like to have to backpedal. But if you play it safe, the problem is that you may be losing out, and I believe you are.

I think it should be obvious where I'm leading. I believe that elementary number theory and the rest of mathematics should be pursued more in the spirit of experimental science, and that you should be willing to adopt new principles. I believe that Euclid's statement that an axiom is a self-evident truth is a big mistake. The Schrödinger equation certainly isn't a self-evident truth! And the Riemann hypothesis isn't self-evident either, but it's very useful.

So I believe that we mathematicians shouldn't ignore incompleteness. It's a safe thing to do but we're losing out on results that we could get. It would be as if physicists said, okay no Schrödinger equation, no Maxwell's equations, we stick with Newton, everything must be deduced from Newton's laws. (Maxwell even tried it. He had a mechanical model of an electromagnetic field. Fortunately they don't teach that in college!)

I proposed all this twenty years ago when I started getting these information-theoretic incompleteness results. But independently a new school on the philosophy of mathematics is emerging called the "quasi-empirical" school of thought regarding the foundations of mathematics. There's a book of Tymoczko's called *New Directions in the Philosophy of Mathematics* (Birkhäuser, Boston, 1986). It's a good collection of articles. Another place to look is *Searching for Certainty* by John Casti (Morrow, New York, 1990) which has a good chapter on mathematics. The last half of the chapter talks about this quasi-empirical view.

By the way, Lakatos, who was one of the people involved in this new movement, happened to be at Cambridge at that time. He'd left Hungary.

The main schools of mathematical philosophy at the beginning of this century were Russell and Whitehead's view that logic was the basis for everything, the formalist school of Hilbert, and an "intuitionist" constructivist school of Brouwer. Some people think that Hilbert believed that mathematics is a meaningless game played with marks of ink on paper. Not so! He just said that to be absolutely clear and precise what mathematics is all about, we have to specify the rules determining whether a proof is correct so precisely that they become mechanical. Nobody who thought that mathematics is meaningless would have been so energetic and done such important work and been such an inspiring leader.

Originally most mathematicians backed Hilbert. Even after Gödel and even more emphatically Turing showed that Hilbert's dream didn't work, in practice mathematicians carried on as before, in Hilbert's spirit. Brouwer's constructivist attitude was mostly considered a nuisance. As for Russell and Whitehead, they had a lot of problems getting all of mathematics from logic. If you get all of mathematics from set theory you discover that it's nice to define the whole numbers in terms of sets (von Neumann worked on this). But then it turns out that there's all kinds of problems with sets. You're not making the natural numbers more solid by basing them on something which is more problematical.

Now everything has gone topsy-turvy. It's gone topsy-turvy, not because of any philosophical argument, not because of Gödel's results or Turing's results or my own incompleteness results. It's gone topsy-turvy for a very simple reason— the computer!

The computer as you all know has changed the way we do everything. The computer has enormously and vastly increased mathematical experience. It's so easy to do calculations, to test many cases, to run experiments on the computer. The computer has so vastly increased mathematical experience, that in order to cope, people are forced to proceed in a more pragmatic fashion. Mathematicians are proceeding more pragmatically, more like experimental scientists do. This new tendency is often called "experimental mathematics." This phrase comes up a lot in the field of chaos, fractals and nonlinear dynamics.

It's often the case that when doing experiments on the computer, numerical experiments with equations, you see that something happens, and you conjecture a result. Of course it's nice if you can prove it. Especially if the proof is short. I'm not sure that a thousand page proof helps too much. But if it's a short proof it's certainly better than not having a proof. And if you have several proofs from different viewpoints, that's very good.

But sometimes you can't find a proof and you can't wait for someone else to find a proof, and you've got to carry on as best you can. So now mathematicians sometimes go ahead with working hypotheses on the basis of the results of computer experiments. Of course if it's physicists doing these computer experiments, then it's certainly okay; they've always relied heavily on experiments. But now even mathematicians sometimes operate in this manner. I believe that there's a new journal called the *Journal of Experimental Mathematics*. They should've put me on their editorial board, because I've been proposing this for twenty years based on my information-theoretic ideas.

So in the end it wasn't Gödel, it wasn't Turing, and it wasn't my results that are making mathematics go in an experimental mathematics direction, in a quasi-empirical direction. The reason that mathematicians are changing their working habits is the computer. I think it's an excellent joke! (It's also funny that of the three old schools of mathematical philosophy, logicist, formalist, and intuitionist, the most neglected was Brouwer, who had a constructivist attitude years before the computer gave a tremendous impulse to constructivism.)

Of course, the mere fact that everybody's doing something doesn't mean that they ought to be. The change in how people are behaving isn't because of Gödel's theorem or Turing's theorems or my theorems, it's because of the computer. But I think that the sequence of work that I've outlined does provide some theoretical justification for what everybody's doing anyway without worrying about the theoretical justification. And I think that the question of how we should actually do mathematics requires **at least** another generation of work. That's basically what I wanted to say—thank you very much!

Bibliography

[1] G. J. Chaitin, *Algorithmic Information Theory*, revised third printing, Cambridge University Press, 1990.

[2] G. J. Chaitin, *Information, Randomness & Incompleteness*, second edition, World Scientific, 1990.

[3] G. J. Chaitin, *Information-Theoretic Incompleteness*, World Scientific, 1992.

[4] G. J. Chaitin, "Exhibiting randomness in arithmetic using *Mathematica* and *C*," *IBM Research Report RC-18946*, 94 pp., June 1993. (To obtain this report in machine readable form, send e-mail to chaitin @ watson.ibm.com.)

PartII
The Berry Paradox

The Berry Paradox

(Originally appeared in Complexity 1, No. 1, (1995))

G. J. Chaitin

IBM Research Division,
P. O. Box 704, Yorktown Heights, NY 10598, USA,
chaitin @ watson.ibm.com

In early 1974, I was visiting the Watson Research Center and I got the idea of calling Gödel on the phone. I picked up the phone and called and Gödel answered the phone. I said, "Professor Gödel, I'm fascinated by your incompleteness theorem. I have a new proof based on the Berry paradox that I'd like to tell you about." Gödel said, "It doesn't matter which paradox you use." He had used a paradox called the liar paradox. I said, "Yes, but this suggests to me an information-theoretic view of incompleteness that I would very much like to tell you about and get your reaction." So Gödel said, "Send me one of your papers. I'll take a look at it. Call me again in a few weeks and I'll see if I give you an appointment."

I had had this idea in 1970, and it was 1974. So far I had only published brief abstracts. Fortunately I had just gotten the galley proofs of my first substantial paper on this subject. I put these in an envelope and mailed them to Gödel.

I called Gödel again and he gave me an appointment! As you can imagine I was delighted. I figured out how to go to Princeton by train. The day arrived and it had snowed and there were a few inches of snow everywhere. This was certainly not going to stop me from visiting Gödel! I was about to leave for the train when the phone rang. It was Gödel's secretary, who said that Gödel was very careful about his health and because of the snow he wasn't coming to the Institute that day. Therefore, my appointment was canceled.

And that's how I had two phone conversations with Gödel but never met him. I never tried again.

I'd like to tell you what I would have told Gödel. What I wanted to tell Gödel is the difference between what you get when you study the limits of mathematics the way Gödel did, using the paradox of the liar, and what I get using the Berry paradox instead.

What is the paradox of the liar? Well, the paradox of the liar is

"This statement is false!"

Why is this a paradox? What does "false" mean? Well, "false" means "does not correspond to reality." This statement says that it is false. If that doesn't correspond to reality, it must mean that the statement is true, right? On the other hand, if the statement is true it means that what it says corresponds to reality. But what it says is that it is false. Therefore the statement must be false. So whether you assume that it's true or false, you must conclude the opposite! So this is the paradox of the liar.

Now let's look at the Berry paradox. First of all, why "Berry"? Well it has nothing to do with fruit! This paradox was published at the beginning of this century by Bertrand Russell. Now there's a famous paradox which is called Russell's paradox and this is not it! This is another paradox that he published. I guess people felt that if you just said the Russell paradox and there were two of them it would be confusing. And Bertrand Russell when he published this paradox had a footnote saying that it was suggested to him by an Oxford University librarian, a Mr G. G. Berry. So it ended up being called the Berry paradox even though it was published by Russell.

Here is a version of the Berry paradox:

> "the first positive integer that cannot
> be specified in less than a billion words".

This is a phrase in English that specifies a particular positive integer. Which positive integer? Well, there are an infinity of positive integers, but at any given time there are only a finite number of words in English. Therefore, if you have a billion words, there's only going to be a finite number of expressions of any given finite length. But there's an infinite number of positive integers. Therefore most positive integers require more than a billion words to describe. So let's just take the first one. But wait a second. By definition this integer is supposed to take a billion words to specify, but I just specified it using much less than a billion words! That's the Berry paradox.

What does one do with these paradoxes? Let's take a look again at the liar paradox:

> "This statement is false!"

The first thing that Gödel does is to change it from "This statement is false" to "This statement is unprovable":

> "This statement is unprovable!"

What do we mean by "unprovable"?

In order to be able to show that mathematical reasoning has limits you've got to say very precisely what the axioms and methods of reasoning are that you have in mind. In other words, you have to specify how mathematics is done with mathematical precision so that it becomes a clear-cut question. Hilbert put it this way: The rules should be so clear, that if somebody gives you what they claim is a proof, there is a mechanical procedure that will check whether the proof is correct or not, whether it obeys the rules or not. This proof-checking algorithm is the heart of this notion of a completely formal axiomatic system.

So "This statement is unprovable" doesn't mean unprovable in a vague way. It means unprovable when you have in mind a specific formal axiomatic system *FAS* with its mechanical proof-checking algorithm. So there is a subscript:

"This statement is unprovable$_{FAS}$!"

And the particular formal axiomatic system that Gödel was interested in dealt with the positive integers and addition and multiplication, that was what it was about. Now what happens with "This statement is unprovable"? Remember the liar paradox:

"This statement is false!"

But here

"This statement is unprovable$_{FAS}$!"

the paradox disappears and we get a theorem. We get incompleteness, in fact. Why?

Consider "This statement is unprovable". There are two possibilities: either it's provable or it's unprovable.

If "This statement is unprovable" turns out to be unprovable within the formal axiomatic system, that means that the formal axiomatic system is incomplete. Because if "This statement is unprovable" is unprovable, then it's a true statement. Then there's something true that's unprovable which means that the system is incomplete. So that would be bad.

What about the other possibility? What if

"This statement is unprovable$_{FAS}$!"

is provable? That's even worse. Because if

"This statement is unprovable$_{FAS}$!"

is provable and it says of itself that it's unprovable, then we're proving something that's false.

So Gödel's incompleteness result is that if you assume that only true theorems are provable, then this

"This statement is unprovable$_{FAS}$!"

is an example of a statement that is true but unprovable.

But wait a second, how can a statement deny that it is provable? In what branch of mathematics does one encounter such statements? Gödel cleverly converts this

"This statement is unprovable$_{FAS}$!"

into an arithmetical statement, a statement that only involves positive integers and addition and multiplication. How does he do this?

The idea is called gödel numbering. We all know that a string of characters can also be thought of as a number. Characters are either 8 or 16 bits in binary. Therefore, a string of N characters is either $8N$ or $16N$ bits, and it is also the base-two notation for a large positive integer. Thus every mathematical statement in a formal axiomatic system is also a number. And a proof, which is a sequence of steps, is also a long character string, and therefore is also a number. Then you can define this very funny numerical relationship between two numbers X and Y, which is that X is the gödel number of a proof of the statement whose gödel number is Y. Thus

"This statement is unprovable$_{FAS}$!"

ends up looking like a very complicated numerical statement.

There is another serious difficulty. How can this statement refer to itself? Well you can't directly put the gödel number of this statement inside this statement; it's too big to fit! But you can do it indirectly. This is how Gödel does it: The statement doesn't refer to itself directly. It says that if you perform a certain procedure to calculate a number, this is the gödel number of a statement which cannot be proved. And it turns out that the number you calculate is precisely the gödel number of the entire statement

"This statement is unprovable$_{FAS}$!"

That is how Gödel proves his incompleteness theorem.

What happens if you start with this

"the first positive integer that cannot
be specified in less than a billion words"

instead? Everything has a rather different flavor. Let's see why.

The first problem we've got here is what does it mean to specify a number using words in English? This is very vague. So instead let's use a computer. Pick a standard general-purpose computer, in other words, pick a universal Turing machine (UTM). Now the way you specify a number is with a computer program. When you run this computer program on your UTM it prints out this number and halts. So a program is said to specify a number, a positive integer, if you start the program running on your standard UTM, and after a finite amount of time it prints out one and only one great big positive integer and it says "I'm finished" and halts.

Now it's not English text measured in words, it's computer programs measured in bits. This is what we get. It's

"the first positive integer that cannot
be specified$_{UTM}$ by a computer program
with less than a billion bits".

By the way the computer program must be self-contained. If it has any data, the data is included in the program as a constant.

Next we have to do what Gödel did when he changed "This statement is false" into "This statement is unprovable." So now it's

> "the first positive integer that can be proved$_{FAS}$
> to have the property that it cannot
> be specified$_{UTM}$ by a computer program
> with less than a billion bits".

And to make things clearer let's replace "a billion bits" by "N bits". So we get:

> "the first positive integer that can be proved$_{FAS}$
> to have the property that it cannot
> be specified$_{UTM}$ by a computer program
> with less than N bits".

The interesting fact is that there is a computer program of length

$$\log_2 N + c_{FAS}$$

bits for calculating this number that supposedly cannot be calculated by any program that is less than N bits long. And

$$\log_2 N + c_{FAS}$$

is much much smaller than N for sufficiently large N. Thus for such N our FAS cannot enable us to exhibit any numbers that require programs more than N bits long. This is my information-theoretic incompleteness result that I wanted to discuss with Gödel.

Why does there have to exist a program that is

$$\log_2 N + c_{FAS}$$

bits long for calculating

> "the first positive integer that can be proved$_{FAS}$
> to have the property that it cannot
> be specified$_{UTM}$ by a computer program
> with less than N bits" ?

Well here is how you show it.

You start running through all possible proofs in the formal axiomatic system in size order. You apply the proof-checking algorithm to each proof. And after filtering out all the invalid proofs, you search for the first proof that a particular positive integer requires at least an N-bit program.

The algorithm that I've just described is very slow but it is very simple. It's basically just the proof-checking algorithm, which is c_{FAS} bits long, and the number N, which is $\log_2 N$ bits long. So the total number of bits is just

$$\log_2 N + c_{FAS},$$

as was claimed. That concludes the proof of my incompleteness result that I wanted to discuss with Gödel.

Over the years, I've continued to do research on my information-theoretic approach to incompleteness. Here are the three most dramatic results that I've obtained thus far:

1) Call a program "elegant" if no smaller program produces the same output. You can't prove that a program is elegant. More precisely, N bits of axioms are needed to be able to prove that an N-bit program is elegant.

2) Consider the binary representation of the halting probability Ω, which is the probability that a program chosen at random halts. You can't prove what one of the bits of Ω is. More precisely, N bits of axioms are needed to be able to determine N bits of Ω.

3) I have constructed a perverse algebraic equation

$$P(K, X, Y, Z, \ldots) = 0.$$

Vary the parameter K and ask whether this equation has finitely or infinitely many whole-number solutions. In each case, this turns out to be equivalent to determining one of the bits of Ω. Therefore N bits of axioms are needed to be able to settle N cases.

These striking examples show that sometimes you have to put more into a set of axioms in order to get more out. Results (2) and (3) are extreme cases. They are accidental mathematical assertions that are true for no reason at all. In other words, the questions considered in (2) and (3) are irreducible; essentially the only way to prove them is to assume them as new axioms. So in this extreme case, what you get out of a set of axioms is only what you put in.

How do I prove these incompleteness results (1), (2) and (3)? As before, the basic idea is the paradox of "the first positive integer that cannot be specified in less than a billion words." For (1) the connection with the Berry paradox is obvious. For (2) and (3) it was obvious to me only in the case where one is talking about determining the **first** N bits of Ω. In the case where the N bits of Ω are scattered about, my original proof of (2) and (3) (the one given in my Cambridge University Press monograph) is decidedly not along the lines of the Berry paradox. But a few years later I was happy to discover a new and more straightforward proof of (2) and (3) that is along the lines of the Berry paradox!

In addition to working on incompleteness, I have also devoted a great deal of thought to the central idea that can be extracted from my version of the Berry paradox, which is to define the program-size complexity of something to be the size in bits of the smallest program that calculates it. I have developed a general theory dealing with program-size complexity that I call algorithmic information theory (*AIT*).

AIT is an elegant theory of complexity, perhaps the most developed of all such theories. But as von Neumann said, pure mathematics is easy compared to the real world! *AIT* provides the correct complexity concept for metamathematics, but not necessarily for other more practical fields.

Program-size complexity in *AIT* is analogous to entropy in statistical mechanics. Just as thermodynamics gives limits on heat engines, *AIT* gives *limits* on formal axiomatic systems.

I have recently reformulated *AIT*. Up to now, the best version of *AIT* studied the size of programs in a computer programming language that was not actually usable. Now I have obtained the correct program-size complexity measure from a powerful and easy to use programming language. This language is a version of *LISP*, and I have written an interpreter for it in *C*. I have written a book employing this new approach that is entitled *The Limits of Mathematics.* To automatically obtain this book in LaTeX, send e-mail to "chao-dyn @ xyz.lanl.gov" with "Subject: get 9407003" or with "Subject: get 9407009". For an extended abstract of the book, request "9407010".

So this is what I would like to discuss with Gödel, if I could speak with him now. Of course this is impossible! But thank you very much for giving me the opportunity to tell you about these ideas!

Questions for Future Research

- Find questions in algebra, topology and geometry that are equivalent to determining bits of Ω.
- What is an interesting or natural mathematical question?
- How often is such a question independent of the usual axioms? (I suspect the answer is "Quite often!")
- Show that a classical open question in number theory, such as the Riemann hypothesis, is independent of the usual axioms. (I suspect that this is often the case, but that it cannot be proven.)
- When doing mathematics, should we take incompleteness seriously or is it a red herring? (I believe that we should take incompleteness very seriously indeed.)
- Is mathematics quasi-empirical? In other words, should mathematics be done more like physics is done? (I believe the answer to both questions is "Yes.")

Bibliography

Books:

- G. J. Chaitin, *Information, Randomness & Incompleteness,* second edition, World Scientific, 1990. Errata: on page 26, line 25, "quickly that" should read "quickly than"; on page 31, line 19, "Here one" should read "Here once"; on page 55, line 17, "RI, p. 35" should read "RI, 1962, p. 35"; on page 85, line 14, "1. The problem" should read "1. The Problem"; on page 88, line 13, "4. What is life?" should read "4. What is Life?"; on page 108, line 13, "the table in" should read "in the table in"; on page 117, Theorem 2.3(q), "$H_C(s,t)$" should read "$H_C(s/t)$"; on page 134, line 7, "$\#\{n|H(n) \leq n\} \leq 2^n$" should read "$\#\{k|H(k) \leq n\} \leq 2^n$"; on page 274, bottom line, "n_{4p+4}" should read "$n_{4p'+4}$".

- G. J. Chaitin, *Algorithmic Information Theory,* fourth printing, Cambridge University Press, 1992. Erratum: on page 111, Theorem I0(q), "$H_C(s,t)$" should read "$H_C(s/t)$".
- G. J. Chaitin, *Information-Theoretic Incompleteness,* World Scientific, 1992. Errata: on page 67, line 25, "are there are" should read "are there"; on page 71, line 17, "that case that" should read "the case that"; on page 75, line 25, "the the" should read "the"; on page 75, line 31, "$-\log_2 p - \log_2 q$" should read "$-p\log_2 p - q\log_2 q$"; on page 95, line 22, "This value of" should read "The value of"; on page 98, line 34, "they way they" should read "the way they"; on page 99, line 16, "exactly same" should read "exactly the same"; on page 124, line 10, "are there are" should read "are there".

Recent Papers:

- G. J. Chaitin, "On the number of n-bit strings with maximum complexity," *Applied Mathematics and Computation* **59** (1993), pp. 97–100.
- G. J. Chaitin, "Randomness in arithmetic and the decline and fall of reductionism in pure mathematics," chao-dyn/9304002, *Bulletin of the European Association for Theoretical Computer Science,* No. 50 (June 1993), pp. 314–328.
- G. J. Chaitin, "The limits of mathematics—Course outline & software," chao-dyn/9312006, *IBM Research Report RC-19324,* 127 pp., December 1993.
- G. J. Chaitin, "Randomness and complexity in pure mathematics," *International Journal of Bifurcation and Chaos* **4** (1994), pp. 3–15.
- G. J. Chaitin, "Responses to 'Theoretical Mathematics...'," *Bulletin of the American Mathematical Society* **30** (1994), pp. 181–182.
- G. J. Chaitin, "The Limits of Mathematics," chao-dyn/9407003, IBM Research Report RC-19646, July 1994, 270 pp.
- G. J. Chaitin, "The Limits of Mathematics IV," chao-dyn/9407009, IBM Research Report RC-19671, July 1994, 231 pp.
- G. J. Chaitin, "The Limits of Mathematics (Extended Abstract)," chao-dyn/9407010, IBM Research Report RC-19672, July 1994, 7 pp.

See Also:

- M. Davis, "What is a computation?," in L.A. Steen, *Mathematics Today,* Springer-Verlag, 1978.
- R. Rucker, *Infinity and the Mind,* Birkhäuser, 1982.
- T. Tymoczko, *New Directions in the Philosophy of Mathematics,* Birkhäuser, 1986.
- R. Rucker, *Mind Tools,* Houghton Mifflin, 1987.
- H.R. Pagels, *The Dreams of Reason,* Simon & Schuster, 1988.
- D. Berlinski, *Black Mischief,* Harcourt Brace Jovanovich, 1988.
- R. Herken, *The Universal Turing Machine,* Oxford University Press, 1988.
- I. Stewart, *Game, Set & Math,* Blackwell, 1989.
- G.S. Boolos and R.C. Jeffrey, *Computability and Logic,* third edition, Cambridge University Press, 1989.

- J. Ford, "What is chaos?," in P. Davies, *The New Physics*, Cambridge University Press, 1989.
- J.L. Casti, *Paradigms Lost*, Morrow, 1989.
- G. Nicolis and I. Prigogine, *Exploring Complexity*, Freeman, 1989.
- B.-O. Küppers, *Information and the Origin of Life*, MIT Press, 1990.
- J.L. Casti, *Searching for Certainty*, Morrow, 1991.
- J.A. Paulos, *Beyond Numeracy*, Knopf, 1991.
- L. Brisson and F.W. Meyerstein, *Inventer L'Univers*, Les Belles Lettres, 1991. (English edition in press)
- J.D. Barrow, *Theories of Everything*, Oxford University Press, 1991.
- D. Ruelle, *Chance and Chaos*, Princeton University Press, 1991.
- T. Nørretranders, *Mærk Verden*, Gyldendal, 1991.
- M. Gardner, *Fractal Music, Hypercards and More*, Freeman, 1992.
- P. Davies, *The Mind of God*, Simon & Schuster, 1992.
- J.D. Barrow, *Pi in the Sky*, Oxford University Press, 1992.
- N. Hall, *The New Scientist Guide to Chaos*, Penguin, 1992.
- H.-C. Reichel and E. Prat de la Riba, *Naturwissenschaft und Weltbild*, Hölder-Pichler-Tempsky, 1992.
- I. Stewart, *The Problems of Mathematics*, Oxford University Press, 1992.
- A.K. Dewdney, *The New Turing Omnibus*, Freeman, 1993.
- A.B. Çambel, *Applied Chaos Theory*, Academic Press, 1993.
- K. Svozil, *Randomness & Undecidability in Physics*, World Scientific, 1993.
- J.L. Casti, *Complexification*, HarperCollins, 1994.
- M. Gell-Mann, *The Quark and the Jaguar*, Freeman, 1994.
- T. Nørretranders, *Verden Vokser*, Aschehdoug, 1994.
- S. Wolfram, *Cellular Automata and Complexity*, Addison-Wesley, 1994.
- C. Calude, *Information and Randomness*, Springer-Verlag, 1994.
- J.L. Casti and A. Karlqvist, *Cooperation and Conflict in General Evolutionary Processes*, Wiley, 1995.

PartIII
Knots and Complex Systems

Knots and Complex Systems

Louis H. Kauffman

Department of Mathematics, Statistics and Computer Science
University of Illinois at Chicago
851 South Morgan Street
Chicago,IL 60607-7045

Abstract. This paper is a survey of topics in knot theory from the point of view of complex systems.

1 Introduction

Knots are fundamental to the topology and geometry of three dimensional spaces. Not only do knots occur in Euclidean three dimensional space, creating the phenomena of linking and tangling of thread and line, but via surgery constructions knots can be used to construct all possible compact three-manifolds. The classification of three manifolds is a special problem in equivalences of knots and links. Diagrammatic calculi such as the Kirby calculus, designed to handle three-manifolds, are just a hair's breadth away from the disordered chaos of all possible tangling. The algebraic structure of knots is related in simple and complex ways with problems in set theory and logic. The structure of topological invariants of knots and links is inextricably tied with the methods and ideas of quantum theory. The range of these relationships in knot theory is very great. The purpose of this paper is to give a flavor of the subject.

The second section discusses the basics of knot theory and an algebraic invariant called the *IQ* (short for involutory quandle) of the knot. The third section discusses braids and magmas. Magmas are algebras with a single binary operation that is distributive over itself. The abstract structure of such algebras is deeply tied with the structure of the Artin braid group. The fourth section discusses the basics of link, twist, writhe and their relations with DNA. Section five is a vignette on the Kirby calculus and the representation of three-manifolds via links. Section six is a quick introduction to quantum link invariants and the ideas behind topological quantum field theory. Section seven is an introduction to knot logic and the representation of non-standard sets via link diagrams. Section eight is a dialogue on the theme of section seven. Section nine is a coda.

It gives the author great pleasure to thank the organizers of the January 1995 - *Conference on Binary Networks and Complex Systems* held in Guana-

juato, Mexico for their wonderful hopitality and encouragement. The contents
of this paper are a rearrangement of the topics covered in the author's talks in
Guanajuato.

2 Reidemeister Moves and the Involutory Quandle

We begin with a description of how the problem of isotopy of knots in three dimensional space is reduced to a combinatorial problem about diagrams drawn in
the plane. This reduction is highly significant. It exchanges one complex problem
for another, as happens so often in topology. However, the combinatorial problem
is quite fascinating, as it lives in a boundary world between geometry/topology
and algebra. As we shall see, this complex world of diagrams is subject to nu-
merous reformulations - allowing a wide range of techniques to impinge on the
theory of knots.

In the beginning we have a single type of spatial move that can be performed
on knots in three dimensional space. This is Reidemeister's *triangle move* as
illustrated in Figure 1.

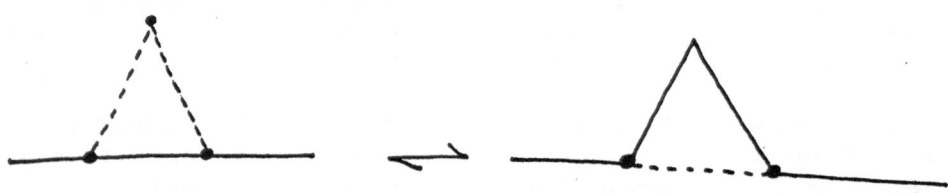

Fig. 1.

In this model all the knots and links in three-space are represented by piecewise linear embeddings. That is, each simple closed curve is represented by a
collection of straight line segments placed end-to-end so that they only touch at
the end points. It is allowed to place a new vertex in the interior of a segment or

to remove a vertex (shared endpoint) if the two segments sharing the endpoint form a straight angle. The triangle move allows take a point in space disjoint from the link and a straight segment on the link. The segment together with the point forms a triangle whose interior surface is stipulated to not intersect the link. Under these conditions the triangle move replaces the original segment by the other two sides of the triangle. Conversely, if two sides of an admissible triangle are given, then they may be replaced by the third side.

In Figure 2 we have illustrated some projections of simple applications of the triangle move.

Next to these projections we have indicated the corresponding moves on knot diagrams. In a knot diagram we do not indicate the piecewise linear structure, but simply draw the arcs as smoothly and simply as possible. The three moves illustrated in Figure 2 are the *Reidemeister moves.* They generate all the other ways of deforming a knot or a link. More precisely, if two links are equivalent in three dimensional space (by some sequence of the three diemsional triangle moves) and if K and K' are diagrams of the links in question, then K' can be obtained from K by a sequence of Reidemeister moves. The converse is obvious via Figure 2. In fact, Figure 2 contains the seeds of the proof of this assertion: Any spatial equivalence can be factored into a sequence of small triangle moves whose projections can be chosen among a small set of types. The main types are shown in Figure 2, and the remaining types can be accomplished by using the moves in Figure 2. For example, view Figure 3 where we show a futher type of projected triangle move and how its results can be obtained through the Reidemeister moves.

Now that we have the Reidemeister moves, lets see what we can do with them. First of all, Figure 4 illustrates an unknotting sequence and also a sequence of moves from the figure eight knot to its mirror image.

Note that each Reidemeister move is performed locally, without disturbing the rest of the diagram, and that we are implicitly using a "move zero" that lets us tidy up a diagram by planar isotopies that do not change the graphical structures of the diagrams. In fact, we need to articulate the move zero as an explicit mathematical operation when we look at link diagrams with respect to a given direction in space.

We shall use the basic set of Reidemeister moves and give the beginnings of a theory of knot invariants. In this theory, we shall assume that each continuous arc in the knot or link diagram is labelled with an element from a "color set" C, and that there is a rule for combining the colors:

$$a, b \to ab$$

that corresponds to the local triplet of arcs that occur at a crossing in the diagram. That is, if c labels the arc you land on by walking along the undercrossing line labelled a, crossing under the arc labelled b, then $c = ab$. See Figure 5.

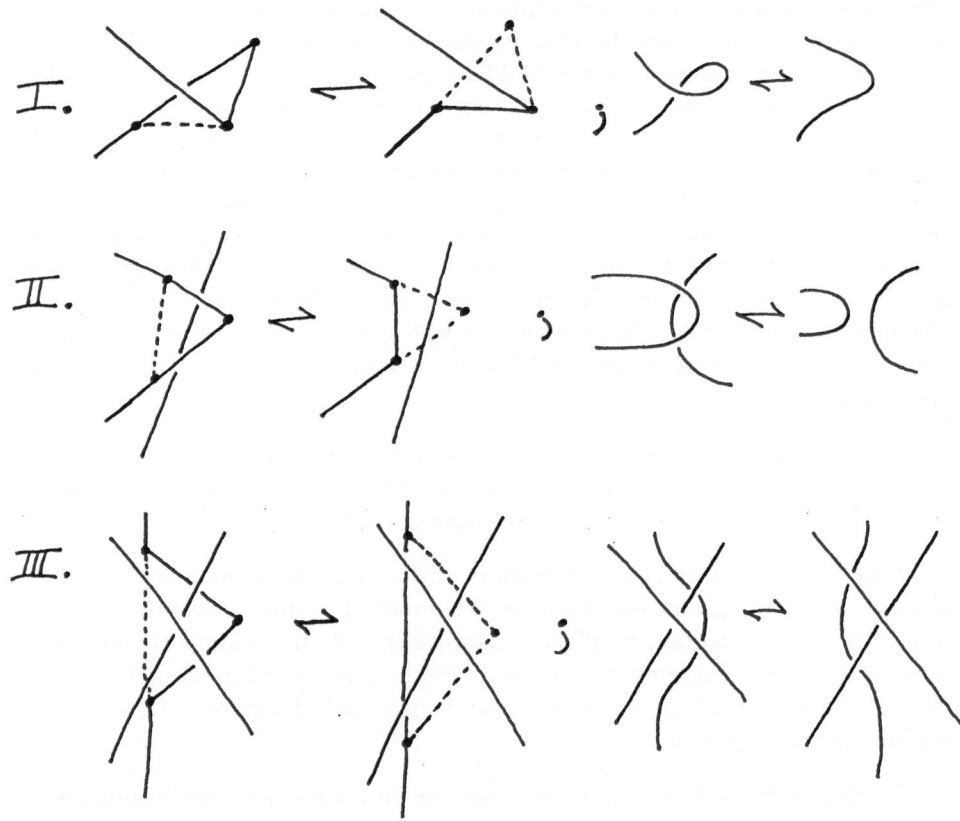

Fig. 2. Reidemeister Moves

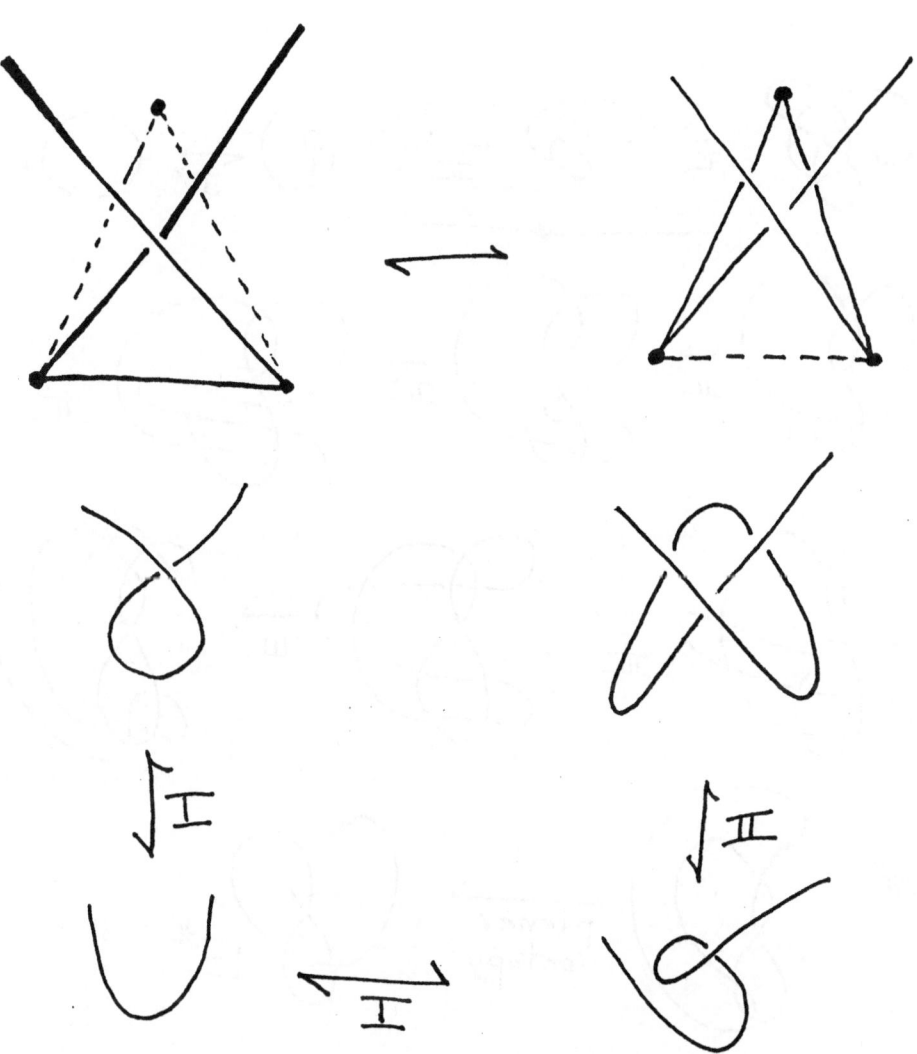

Fig. 3.

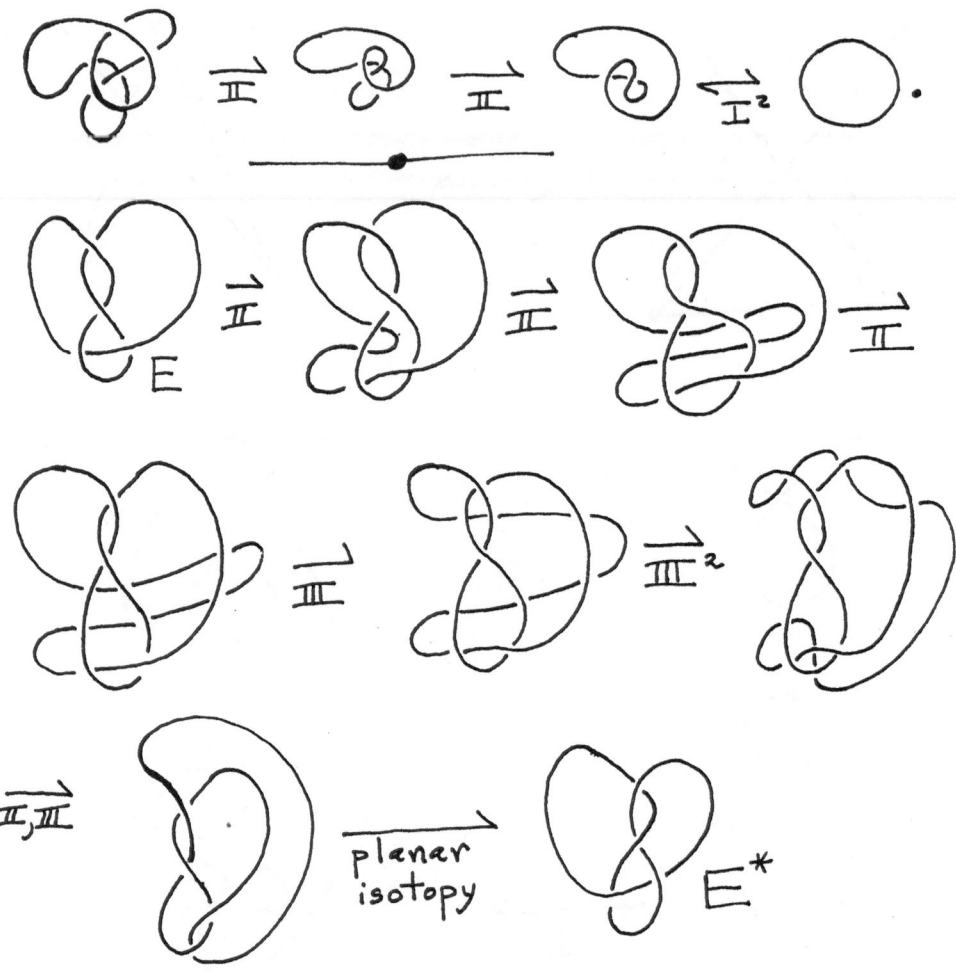

Fig. 4.

In order for this rule of composition to be well defined it must be possible to consider walking back along the arc labelled c, under the arc labelled b, to the arc labelled a. This implies that $a = cb$. Thus we have

$$a = (ab)b$$

by substituting ab for c. This rule is a neccessary axiom for our labelling system. In fact, it corresponds to the second Reidemeister move! For consider the diagram in Figure 5. Here we see an arc labelled a underpassing an arc labelled b *twice* because the b arc doubles back on itself in the form of a second Reidemeister move. As a result the arc that emerges from the double-back is labelled $(ab)b$ and this is equal to just a. As a result, we conclude that a diagram that has been colored with the rules we have established so far has the property that its coloration can be extended (by making only local changes) to a coloration of any diagram obtained from the given one by a second Reidemeister move.

With this idea in mind we can analyse the corresponding conditions for the first and the third Reidemeister moves. The result for the first move is $aa = a$ and the result for the third move is $(ab)c = (ac)(bc)$. These facts are illustrated in Figure 6.

Thus we have found the following requirements for a "good" color set for a knot or link diagram: The color set should have a binary operation on it, denoted ab, for a and b in the color set. For all a, b, c in the color set C, the following equations must hold

$$aa = a,$$

$$(ab)b = b,$$

$$(ab)c = (ac)(bc).$$

A set C satisfing these axioms is called an *involutory quandle*. The terminology "quandle" is due to David Joyce [8].

Given a knot or link diagram K, we can form a universal involutory quandle $IQ(K)$ (denoted the *i-quandle of K* by putting one abstract color label on each arc of the diagram and demanding the relations of the form $c = ab$ at each crossing. The resulting system of labels and all the (non-associative) products of these labels are then assumed to satisfy the axioms for a quandle. In other words, $IQ(K)$ is the quotient by the relations generated by the crossings of the free i-quandle generated by the arc labels in the diagram. It is clear that if K' is obtained from K by a sequence of Reidemeister moves, then $IQ(K')$ is isomorphic (as an algebra) to $IQ(K)$. Thus the i-quandle of a knot or link is a topological invariant of that knot or link.

Note that in forming this universal i-quandle, IQ(K), we have allowed the possiblility of multiplying any two elements of the color set, whether or not they occur together at a crossing in a given diagram of the knot or link. This provides the convenience of abstract algebra at the expense of a lack of direct connection with the geometry of the diagrams. The i-quandle IQ(K) is not a complete

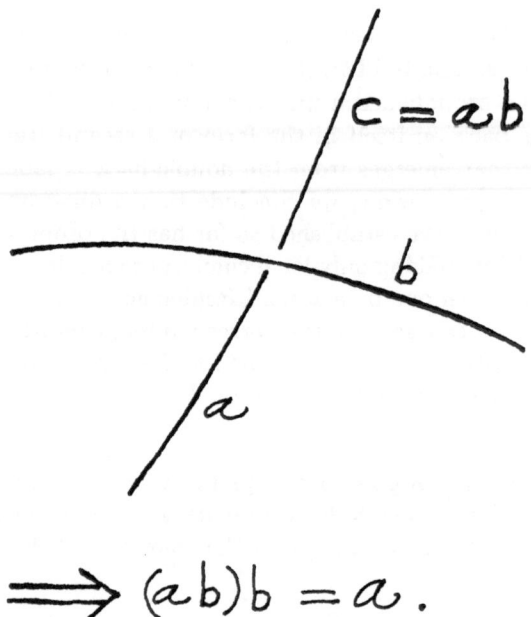

$$\Longrightarrow (ab)b = a.$$

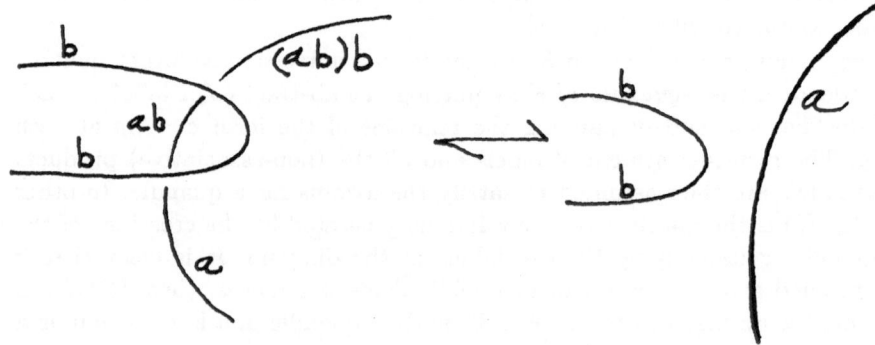

Fig. 5.

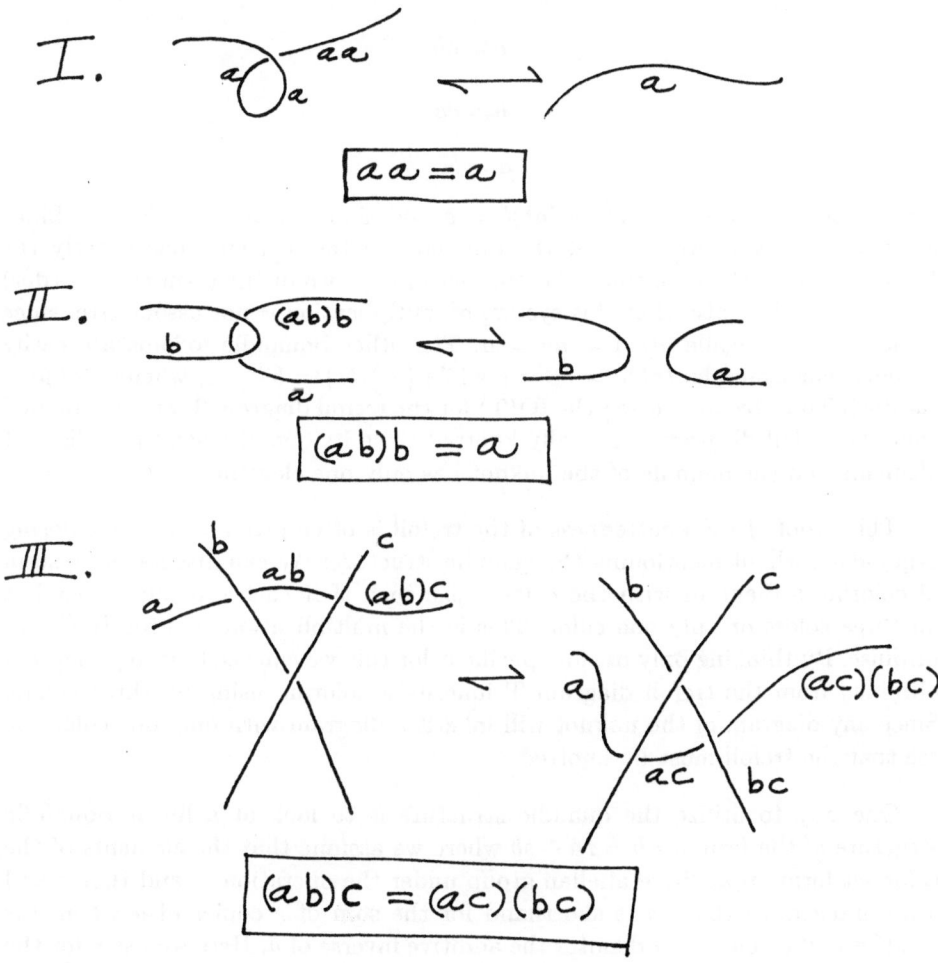

Fig. 6.

invariant of knots. There are inequivalent diagrams that have isomorphic quandles. The advantage of a less than perfect invariant lies in its computability and
simplicity of structure. IQ(K) is sometimes a finite system. There is an oriented
version of IQ(K) that we shall mention shortly. The oriented version is very
nearly a complete knot invariant, and consequently it is a tougher nut to deal
with on the algebraic level.

In Figure 7 we have illlustrated the formation of the i-quandle of the trefoil
diagram T. Here the relations are

$$c = ab$$

$$b = ca$$

$$a = bc.$$

Note that $c = ab$ implies $cb = (ab)b = a$ and similarly $ba = c$, $ac = b$. Since
$xx = x$ for any x, we see that the i-quandle of the of trefoil has exactly the
three elements a,b and c since the product of any two of them (in either order)
is the third. Note also that this system of mutiplication is not associative, since
$(ab)c = cc = c$ while $a(bc) = aa = a$. The other i-quandle axioms are easily
verified. For example, $(ab)c = cc = c$ while $(ac)(bc) = ba = c$, whence $(ab)c =
(ac)(bc)$. Thus, by computing the IQ(T) for the trefoil diagram T we have proved
that the trefoil diagram is actually knotted since its quandle has three distinct
elements and the quandle of the unknot has only one element.

This proof of the knottedness of the trefoil is often given as a pure coloring
argument without mentioning the quandle structure: We can discuss the system
of coloring a diagram with the colors a,b,c such that each crossing either has
all three colors or only one color. This is the multiplication rule for $IQ(T)$ in
disguise. By thinking only of this specific color rule we can see that any diagram
obtained from the trefoil diagram T inherits a coloring using all three colors.
Since any diagram of the unknot will inherit a diagram with only one color, we
see that the trefoil must be knotted.

One way to utilize the quandle structure is to look at a linear i-quandle
structure of the form $a * b = ra + sb$ where we assume that the elements of the
color set form an additive abelian group under the operation $+$ and that r and
s are integers so that ra is shorthand for the sum of r copies of a when r is
positive and $(-1)a = -a$ denotes the additive inverse of a. Here we use $*$ for the
quandle operation in order to distinguish it from the operations in the abelian
group. What does it take for $a * b = ra + sb$ to be a quandle operation? Lets try
the first axiom: $a = a * a = ra + sa = (r + s)a$. This will be satisfied if $r + s = 1$.
Now take the second axiom: $a = (a * b) * b = r(ra + sb) + sb = r^2a + (r + 1)sb$.
This will be satisfied if $r = -1$. Thus we can try $r = -1$ and $s = 2$ so that
$r + s = 1$. In other words, the first two axioms are satisfied for

$$a * b = 2b - a.$$

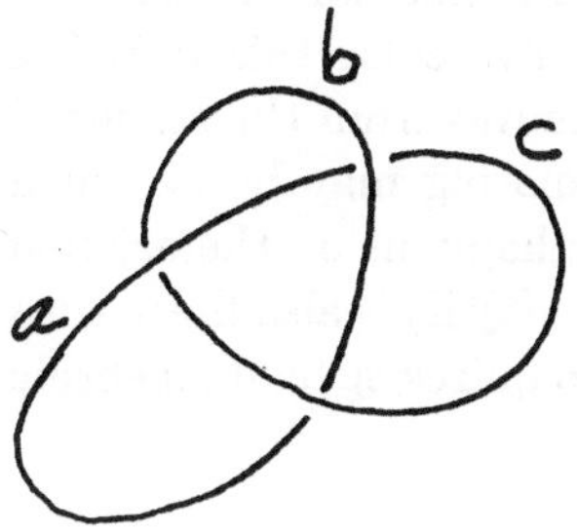

$$c = ab$$
$$a = bc$$
$$b = ca$$
$$\Rightarrow$$
$$cb = a$$
$$ac = b$$
$$ba = c$$

	a	b	c
a	a	c	b
b	c	b	a
c	b	a	c

Fig. 7.

In fact this yields the third axiom as well via $(a * b) * c = 2c - (2b - a) = 2(2c - b) - (2c - a) = (ac) * (bc)$.

With the coloring rule $a * b = 2b - a$ in place we can proceed to determine specific quandles for given knots and links by starting with integer labels on the arcs and finding the restriction imposed by the relations in the knot. For example, in the case of the trefoil we know that $c = a * b, b = c * a, a = b * c$. Hence, $a = (c * a) * c = 2c - (2a - c) = 3c - 2a$ and therefore $3(a - c) = 0$. We conclude that in using integer labelling it is neccessary to work in the integers modulo three in order to produce a trefoil labelling. In this case the structure gives rise to the same 3-element i-quandle that we derived abstractly. In general, every knot has an associated modulus (the least non-zero modulus for a modular number system that will give a consistent quandle structure of this kind) and the modulus is an invariant of the knot. This invariant is often called the *determinant* of the knot [6].

The general quandle structure is of more interest than any particular coloring rule since it gives a way to make an image of the topology in algebra and it gives a way to discover coloring rules that can work for specific knots and links. The next simplest knot, the figure eight knot with diagram E as shown in Figure 8, is a good case in point. Using the rule $a * b = 2b - a$ it is not hard to discover that the modulus of the figure eight knot is five. The five color rule uses five colors to color any diagram obtained by Reidemeister moves from the standard diagram shown in Figure 8. In any given diagram a coloring may be obtained with less than the requisite five colors. In this case that happens on the original diagram E. For the figure eight knot the abstract quandle IQ(E) is also finite with five elements. In general, IQ(K) can be infinite and it requires special algebraic techniques to determine when IQ(K) is finite.

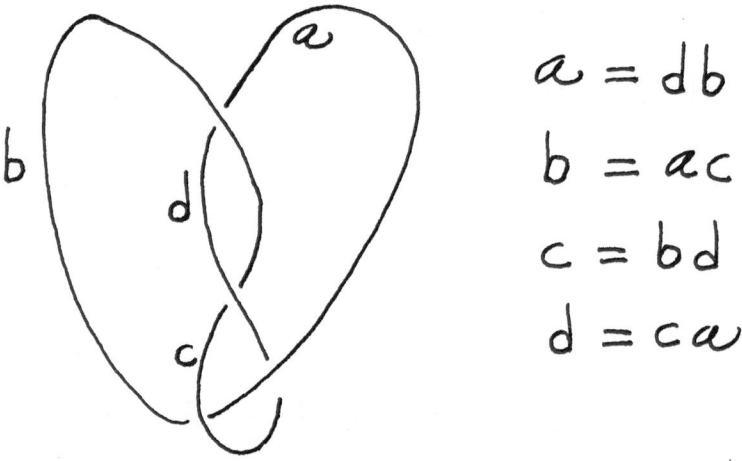

Fig. 8.

Subtle techniques due to Steve Winker [20],[14] allow us to look directly at the structure of IQ(K) for many K and to determine finiteness if it occurs. The basic observation for this technique is the following Lemma. In stating this lemma we use the convention that *an unparenthesized product is canonically left associated.* Thus

$$(xy)z = xyz$$

and

$$abcdefg = (((((ab)c)d)e)f)g.$$

Lemma 1. *In any i-quandle Q,*

$$a(bc) = acbc$$

for any elements a,b,c in Q.

Proof.

$$acbc = ((ac)b)c = ((ac)c)(bc) = a(bc).$$

This completes the proof.

The strategy is to use this rewrite Lemma to simplify the relations in the quandle until they are manageable through a combination of algebra and diagrammatics. Here is an example for the figure eight diagram E. As in Figure 8, we know that $IQ(E)$ is described by the relations

$$a = db$$
$$b = ac$$
$$c = bd$$
$$d = ca.$$

Eliminating d and using the Lemma, we find

$$a = cab$$
$$b = ac$$
$$c = b(ca) = baca.$$

Now eliminating b and using the Lemma, we find

$$a = ca(ac) = cacac$$
$$c = (ac)aca = acaca.$$

Using the second axiom it is easy to see that these two relations imply each other. Thus the unoriented quandle for the figure eight knot E is described by a single relation

$$a = cacac.$$

It follows directly from this that the quandle has the five elements $a,c,ca,cac,caca$.

The first knot with an infinite involutory quandle is the knot 8_{16} on the Reidemeister tables [20]. See Figure 9 for an illustration of 8_{16}.

We should also mention that $IQ(K)$ is a pretty smart quandle. In particular, if K is a non-trivial knot, then $IQ(K)$ is a non-trivial algebraic system [20] and hence IQ *detects knottedness.* Of course it is not neccessarily easy to to detect the non-triviality of an algebra that is given by axioms, generators and relations. Thus this result does not give a direct algorithm for detecting knots.

Fig. 9.

Remark. There is an oriented version of the involutory quandle. This is called a *quandle* and it has two binary operations corresponding to the knot theoretic possibilities of encountering a crossing that goes to the right or to the left. Elucidation of this issue leads to many significant structures in the knot theory. See [11] and [4].

3 Braids and Magmas

The notion of studying a system with a binary orperation that distributes over itself has occurred in other contexts. A striking example of this phenomenon is the investigation of *magmas*, freely generated algebras with a self-distributive law and no other axioms. In [3] the history of this subject is recounted in relation to set theory and the consequences of axioms asserting the existence of very large cardinals. Many properties of magmas were deduced from such axioms before

DeHornoy discovered a way to embed a magma in the infinite Artin braid group. The key to this embedding is the basic relation of self-distributivity and the third Reidemeister move, as we have described it, and the details are quite subtle, leading to new proofs of properties of magams and new proofs and algorithms in the theory of braids.

We now indicate how the concepts for magmas appear after a change of notation. In magma literature, the algebra is *left distrbutive* over itself rather than right distributive. Thus we have $a(bc) = (ab)(ac)$ as the sole axiom of a magma. It turns out to be convenient to use a "Polish" notation for the binary operation, writing $a[b]$ for ab. This means that $(ab)c = a[b][c]$ while $a(bc) = a[b[c]]$. The left-disributive law then becomes

$$a[b[c]] = a[b][a[c]].$$

Lets rewrite this equation with A capitalized:

$$A[b[c]] = A[b][A[c]].$$

In this form the equation states that A is a structure preserving mapping where the structure that is being preserved is the "composition" $b[c]$. Thus A applied to the composition of b and c results in the composition of $A[b]$ and $A[c]$. It is the consideration of structure preserving mappings of cardinals that gives rise to the magma in the set theory context. Axioms about large cardinals ensure the existence of structure preserving mappings of cardinals (called *ranks*) that are not equal to the identity.

In order to describe Dehornoy's construction, we must first recall the structure of the braid group. We regard the Artin braid group B_∞ as the union of braid groups B_n on n strands where B_n is embedded in B_{n+1} by adding a trivial $n+1$-st strand on the right. Then braid group B_n is generated by the elementary braids $\sigma_1, ..., \sigma_{n-1}$ and their inverses where σ_i is a braid where only the i-th and $i + 1$-th strands cross as shown in Figure 10.

In general, a braid in B_n is a configuration of n strands in a plane crossed with the unit interval, so that the strands have a specific row of starting points in the top plane and a corresponding row of ending points in the bottom plane. Each planar cross section of the strands consists in n points. Thus each strand descends from the top plane to the bottom plane, possibly winding about its neighbors. B_n becomes a group through the composition induced by attaching the bottom points of one braid to the top points of the other. The inverse of a braid is its mirror image obtained by reversing all the crossings in a planar projection of the braid. The group B_n is generated by $\sigma_1, ..., \sigma_{n-1}$ and has a complete list of relations:

$$\sigma_i \sigma_{i+1} \sigma_i = \sigma_{i+1} \sigma_i \sigma_{i+1}$$

and

$$\sigma_i \sigma_j = \sigma_j \sigma_i$$

for $|i - j| \geq 2$.

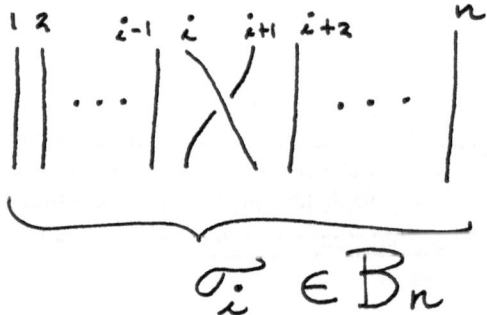

Fig. 10.

The first relation is a version of the third Reidemeister move. The fact that

$$\sigma_i \sigma_i^{-1} = 1$$

is an expression of the second Reidemeister move.

Let M denote the free left-distributive magma on one generator a. Let B_∞ denote the infinite Artin braid group. Dehornoy's construction takes elements of the magma into B_∞. Let X be an element of the magma and $b(X)$ its corresponding braid. Then $b(X)$ is defined inductively by the formulas

$$b(a) = 1$$

and

$$b(X[Y]) = b(X)s(b(Y))\sigma_1 s(b(X)^{-1}).$$

Here $s(b)$ is the braid obtained from the braid b by shifting all its strands to the right by one strand (add one straight strand to the left of b) and $b(X)^{-1}$ denotes the inverse of $b(X)$ in the braid group.

We have constructed the mapping

$$b : M[a] \longrightarrow B_\infty,$$

and leave as an exercise for the reader to check that

$$b(X[Y[Z]]) = b(X[Y][X[Z]])$$

for all X, Y, Z in $M[a]$. This construction embeds the magma into the infinite braid group and allows the solution of the word problem in the braid group to be applied to give a solution of the word problem in the magma. Here we see an application of topology to a problem arising from logic and set theory. This relationship is the tip of an iceberg.

4 Link, Twist,Writhe and DNA

In many applications it is useful to regard a knot as an embedded band in three dimensional space. This is called a "framed knot". It is assumed that the band is topologically equivalent to an annulus, that is it has two boundary curves and it is an orientable surface. The band keeps track of twisting as the knot is deformed. In order to specify a framed knot we need to give the embedding of its core (the curve running along the center of the band) and the structure of the twisting in the band. Note that a "curl" and a "full twist" interchange under ambient isotopy as shown in Figure 11. Thus in studying framed knots the first Reidemeister move is no longer present. In its place we have the exchange of curls and twists.

Fig. 11.

It is not always so obvious just how much linking there is in a given picture of of a framed knot. Lets formalize this concept. We need the notion of *linking number*. Given a link of two components A and B with an orientation assigned to each component, we wish to define an integer $Lk(A, B)$, the linking number of A with B. In terms of the combinatorics of link diagrams we can define the linking number by the formula

$$Lk(A, B) = \sum_{c \in CR(A,B)} sgn(c)$$

where $CR(A, B)$ denotes the set of crossings that A and B have in common, and the sign of a crossing c, denoted $sgn(c)$, is $+1$ or -1 according to the local orientations of the arcs at the crossing as shown in Figure 12.

In Figure 12 we have shown how this definition of linking number gives the answer "one" for the simple link of two circles. This is the "right" answer. Furthermore, it is easy to see the the linking number does not change under the Reidemeister moves. Thus linking number is indeed a link invariant and an easy one to define at that.

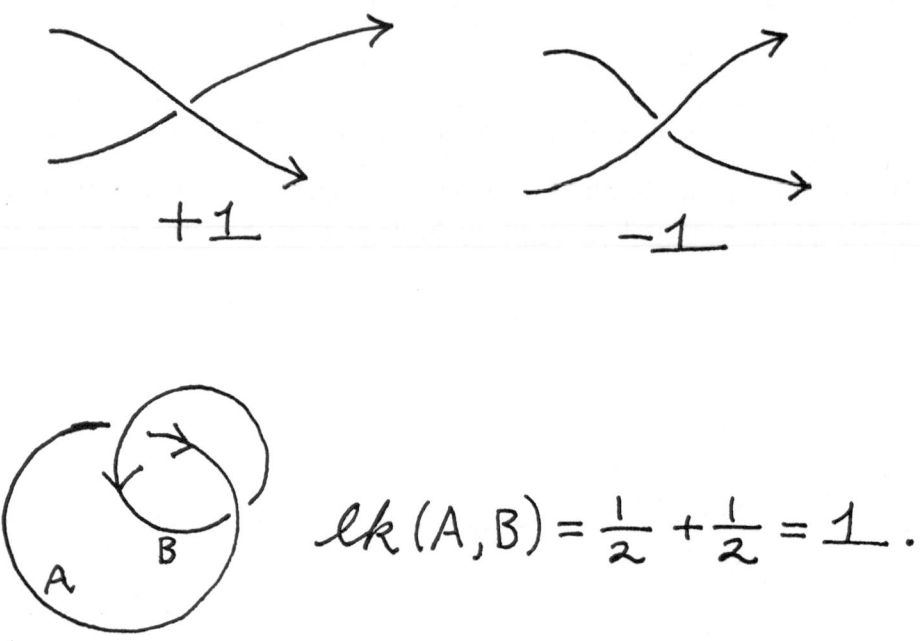

Fig. 12.

Now lets turn to the following question: *Given a framed knot, represented as an embedded band in diagrammatic form, what is the linking number of the two curves that form the boundary of the band?* Here it is assumed that the two curves are given parallel orientations.

It is not hard to see that we can assume that the diagram for the framed link has the form shown in Figure 13. That is, we shall assume that the bands cross each other transversely, producing four crossings of arcs per one crossing of bands, and that the bands have some twists in them - isolated from the crossings of the bands.

With this representation in place, we can count the linking number by breaking the count into the contribution from the crossings of the bands and the contribution from the twists in the bands. It is easy to see that each band crossing contributes exactly plus one or minus one to the linking number according to the sign of the corresponding crossing of the core curve. Thus we define the *writhe*, $w(K)$, of a given knot diagram K to be the sum of the signs of the crossings of

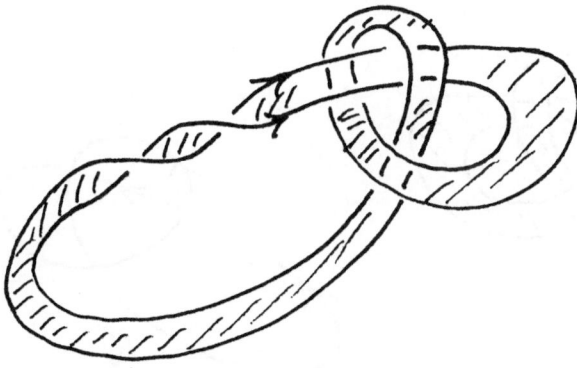

Fig. 13.

K. If B is a framed knot with core K, then

$$Lk(bd(B)) = w(K) + Tw(B)$$

where $bd(B)$ denotes the two curves forming the boundary of B, and Tw(B) denotes the number of twists in the band B that can be counted directly away from its crossings. This formula is a special case of the differential geometric formula of White [19], and it has many applications in molecular biology. In the molecular biology application the knotted band consists of two interwound strands of DNA. In Figure 14 we show some examples of linking number calculations using this formula.

Remark. **Regular Isotopy and Ribbon Equivalence**

We say that two links are *regularly isotopic* if one can be obtained from the other by a sequence of Reidemeister moves of type II or type III. The first move is not allowed in regualr isotopy. Any regular isotopy class of links determines a framed link by thickening it as shown in Figure 15.

Regularly isotopic links give rise to ambient isotopic framed links. In fact, if we add the equivalence shown in Figure 15, and call this *ribbon equivalence* of links, then ribbon equivalence classes of link diagrams are in one-to-one correspondence with ambient isotopy classes of framed links. Ribbon equivalence classes of link diagrams are quite useful in keeping track of the intricacies of DNA strands and they are also useful in working with three dimensional manifolds.

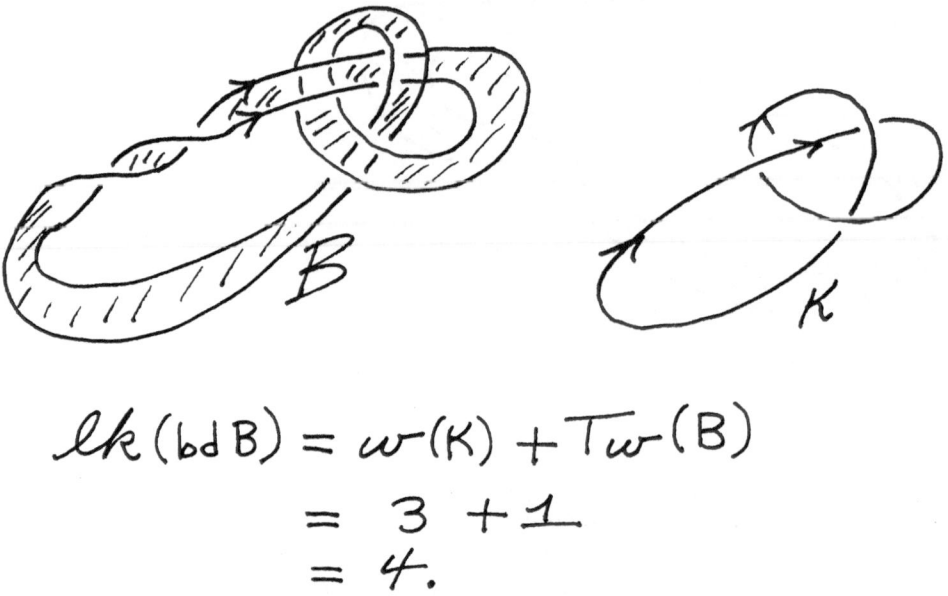

$$\ell k\,(bd\,B) = \omega(K) + Tw(B)$$
$$= 3 + 1$$
$$= 4.$$

Fig. 14.

5 Kirby Calculus

Link diagrams up to ribbon equivalence encode three dimensional manifolds. The correspondence from links to three-manifolds goes via surgery: For each link component we excise a tubular neighborhood and draw a longitude λ along the boundary of the tubular neighborhood in exactly the pattern of the diagram of this component. This lets the diagram of the link component specify both the tubular neighborhood and the choice of longitude. Then a new solid torus is sewn to each tubular neighborhood boundary in such a way that the standard meridian (bounding a disk in the new solid torus) as attached to the longitude on the neighborhood and the standard longitude on the new solid torus is attached to the standard meridian of the neighborhood. The manifold that results form this surgery process associated with the link diagram is denoted $M^3(K)$ and is called the three-manifold obtained via surgery on K.

Every compact three-manifold without boundary can be obtained by this process of surgery on a link [16] and there is a special set of moves on link diagrams that corresponds to equivalences of three manifolds . These moves are listed in the Figure 16 below.

The first move entails the adddition or removal of an unknotted curve with writhe plus one or minus one. This is called *blowing up* when the curve is added

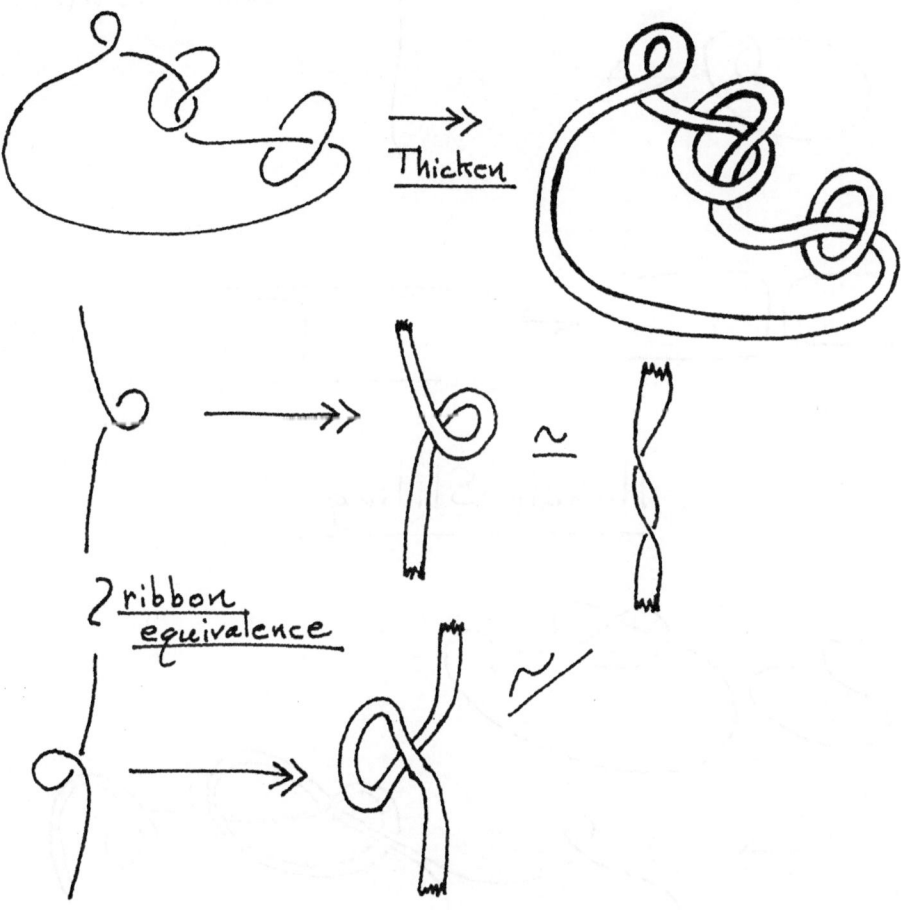

Fig. 15.

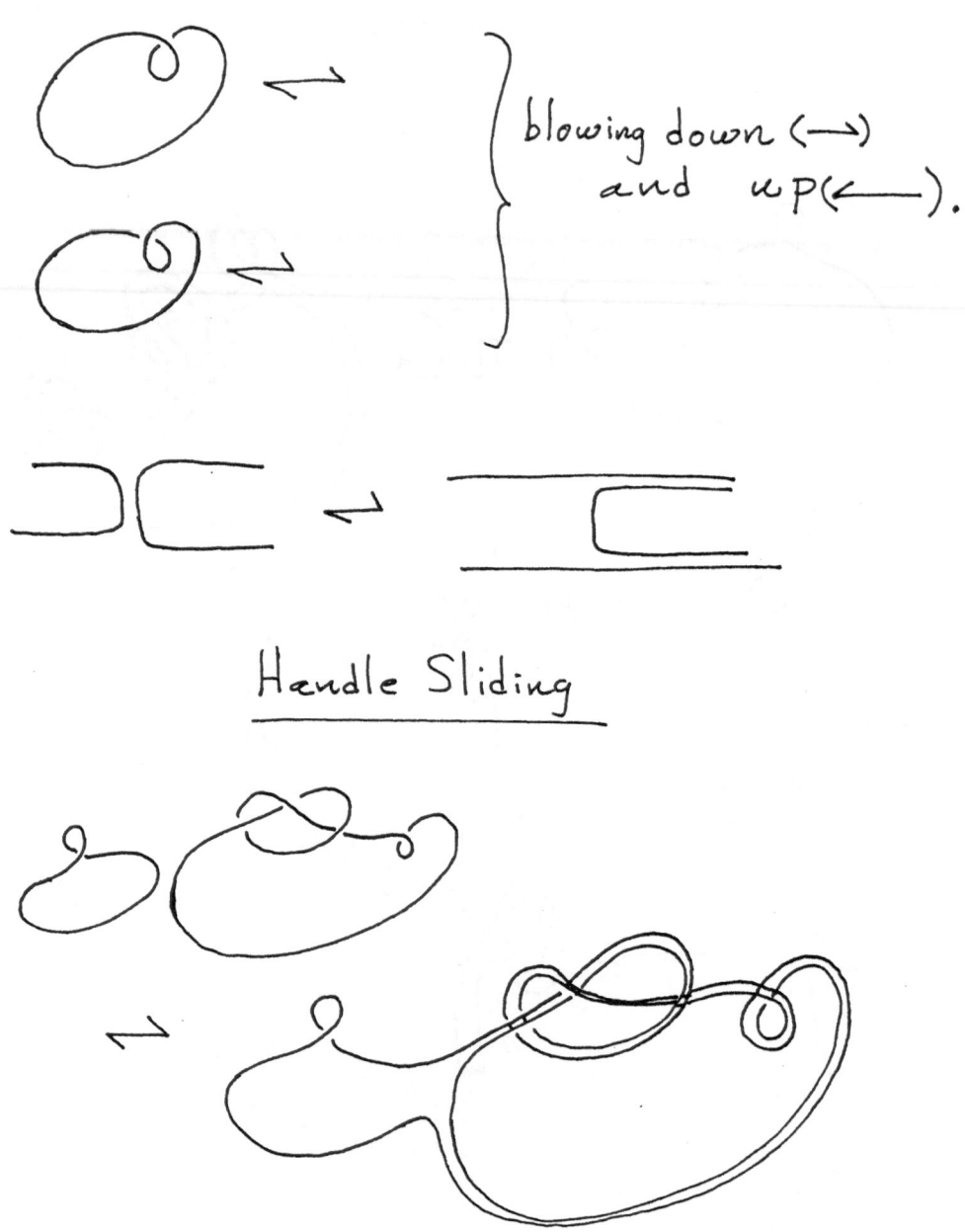

blowing down ($\longrightarrow$)
and up ($\longleftarrow$).

Handle Sliding

Fig. 16. Kirby Calculus

and *blowing down* when the curve is subtracted. This move works because surgery on such a curve has no effect on the topology. The second move is called *handle-sliding* and consists in the creation of a parallel copy of a component and its connection (by a connected sum) with another component. This move creates a radical change in the link diagram but does not effect the three manifold that results from the surgery. The move can be interpreted in terms of sliding handles on a four-manifold whose boundary is $M^3(K)$ and this is how it acquires its name. Kirby [15] showed that *two three manifolds $M^3(K)$ and $M^3(K')$ are homeomorphic if and only if there is a sequence of blowing up and down and handle slides, combined with ribbon equivalence, taking one link K to the other link K'.* Thus Kirby's theorem reduces the problem of classification of three-manifolds to the study of a formal system involving link diagrams.

6 Quantum Link Invariants

There is a remarkable connection between low-dimensional topology and basic ideas in quantum theory.

6.1 Dirac Brackets

Recall Dirac notation, $< a|b >$. In this notation $< a|$ and $|b >$ are vectors and covectors respectively. $< a|b >$ is the evaluation of $< a|$ by $|b >$, hence it is a scalar, and in ordinary quantum mechanics it is a complex number. One can think of this as the amplitude for the state to begin in RaS and end in RbS. That is, there is a process that can mediate a transition from state a to state b.

Except for the fact that amplitudes are complex valued, they obey the usual laws of probability. This means that if the process can be factored into a set of intermediate states $c_1, c_2, ..., c_n$ so that we have the set of processes

$$a \to c_i \to b$$

for $i = 1, ..., n$, then the amplitude for $a \to b$ is the sum of the amplitudes for $a \to c_i \to b$. Meanwhile, the amplitude for $a \to c_i \to b$ is the product of the amplitudes of the two subconfigurations $a \to c_i$ and $c_i \to b$.

Formally we have

$$< a|b >= \sum_i < a|c_i >< c_i|b >$$

where the summation is over all the intermediate states $i = 1, ..., n$.

In general, the amplitude for mutually disjoint processes is the sum of the amplitudes of the individual processes. The amplitude for a configuration of disjoint processes is the product of their individual amplitudes.

Dirac's division of the amplitudes into bras $< a|$ and kets $|b >$ is done mathematically by taking a vector space V (a Hilbert space, but it can be finite dimensional) for the bras: $< a|$ belongs to V. The dual space V^* is the home of the kets. Thus $|b >$ belongs to V^* so that $|b >$ is a linear mapping $|b >: V \to C$ where C denotes the complex numbers. We restore symmetry to the definition by realizing that an element of a vector space V can be regarded as a mapping from the complex numbers to V. Given $< a| : C \to V$, the corresponding element of V is the image of 1 (in C) under this mapping. In other words, $< a|(1)$ is a member of V. Now we have $< a| : C \to V$ and $b >: V \to C$. The composition $< a||b >=< a|b >: C \to C$ is regarded as an element of C by taking $< a|b > (1)$. The complex numbers are regarded as the "vacuum", and the entire amplitude $< a|b >$ is a "vacuum to vacuum" amplitude for a process that includes the creation of the state a, its transition to b, and the annihilation of b to the vacuum once more.

6.2 Knot Amplitudes

At this point a rich imagery arises that goes beyond quantum mechanics – into modern knot theory. Consider first a circle in a spacetime plane with time represented vertically and space horizontally (Figure 17).

The circle represents a vacuum to vacuum process that includes the creation of two "particles" (Figure 18).

and their subsequent annihilation (Figure 19).

In accord with our previous description, we could divide the circle into these two parts (creation "cup" and annihilation "cap") and consider the amplitude $< cup|cap >$. Since the diagram for the creation of the two particles ends in two separate points, it is natural to take a vector space of the form $V \otimes V$ as the target for the bra and as the domain of the ket. We imagine at least one particle property being catalogued by each factor of the tensor product. For example, a basis of V could enumerate the spins of the created particles.

The first hint of topology comes when we realize that it is possible to draw a much more complicated simple closed curve in the plane that is nevertheless decomposed with respect to the vertical direction into many cups and caps. In fact, any non-self-intersecting differentiable curve can be rigidly rotated until it is in general position with respect to the vertical. It will then be seen to be decomposed into these minima and maxima. Our prescriptions for amplitudes suggest that we regard any such curve as an amplitude.

Each simple closed curve gives rise to an amplitude, but any simple closed curve in the plane is isotopic to a circle, by the Jordan Curve Theorem. If these are topological amplitudes, then they should all be equal to the original amplitude for the circle. Thus the question: What condition on creation and annihilation will insure topological amplitudes? The answer derives from the fact

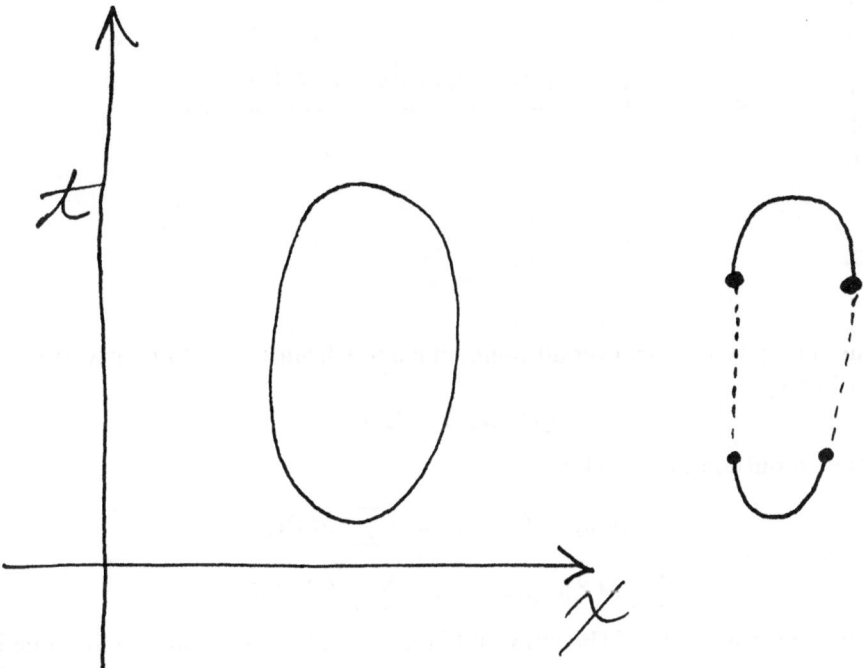

Fig. 17.

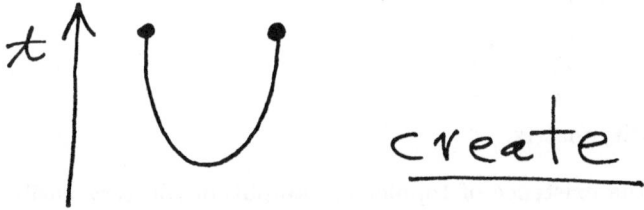

create

Fig. 18.

that all isotopies of the simple closed curves are generated by the cancellation of adjacent maxima and minima as illustrated in Figure 20.

This condition is said very simply by taking a matrix representation for the corresponding operators.

Specifically, let $\{e_1, e_2, ..., e_n\}$ be a basis for V. Let $e_{ab} = e_a \otimes e_b$ denote the elements of the tensor basis for $V \otimes V$. Then there are matrices M_{ab} and M^{ab} such that

$$< cup|(1) = \sum M^{ab} e_{ab}$$

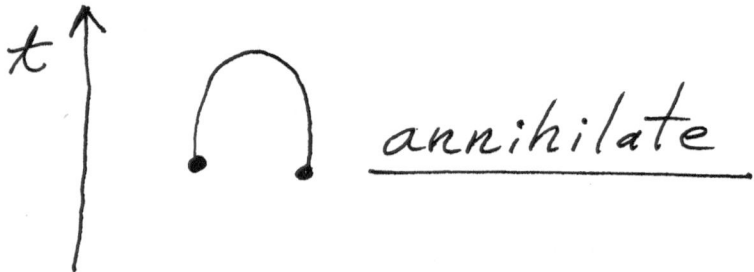

Fig. 19.

with the summation taken over all values of a and b from 1 to n. Similarly, $|cap >$ is described by

$$|cap > (e_{ab}) = M_{ab}.$$

Thus the amplitude for the circle is

$$< cup|cap > (1) = |cap > \sum M^{ab} e_{ab}$$

$$= \sum M^{ab} |cap > (e_{ab}) = \sum M^{ab} M ab.$$

In general, the value of the amplitude on a simple closed curve is obtained by translating it into an Rabstract tensor expressionS in the M^{ab} and M_{ab}, and then summing over these products for all cases of repeated indices.

Returning to the topological conditions we see that they are just that the matrices M^{ab} and M_{cd} are inverses in the sense that

$$\sum_i M^{ai} M_{ib} = \delta_b^a$$

where δ_b^a denotes the identity matrix.

Thus the problem of the existence of topological amplitudes is very easily solved for simple closed curves in the plane.

Now we go to knots and links. Any knot or link can be represented by a picture that is configured with respect to a vertical direction in the plane. The picture will decompose into minima (creations) maxima (annihilations) and crossings of the two types shown in Figure 21. (Here I consider knots and links that are unoriented. They do not have an intrinsic preferred direction of travel.)

Next to each of the crossings we have indicated mappings of $V \otimes V$ to itself, called R and R^{-1} respectively. These mappings represent the transitions corresponding to elementary braiding. We now have the vocabulary of cup, cap, R and R^{-1}. Any knot or link can be written as a composition of these fragments, and consequently a choice of such mappings determines an amplitude for knots and links. In order for such an amplitude to be topological (i.e. an invariant of framed links) we want it to be invariant under a list of local moves on the diagrams. These moves are an augmented list of the Reidemeister moves, adjusted to take care of the fact that the diagrams are arranged with respect to a given direction in the plane. The moves translate into very interesting algebra. Move 3. is the famous Yang-Baxter equation that occurred for the first time in problems of exactly solved models in statistical mechanics. All the moves taken together are directly related to the axioms for a quasi-triangular Hopf algebra (aka quantum group). We shall not go into this connection here.

There is an intimate connection between knot invariants and the structure of generalized amplitudes. The strategy for the construction of such invariants is directly motivated by the concept of an amplitude in quantum mechanics. It turns out that the invariants that can actually be produced by this means are incredibly rich. They encompass, at present, all of the known invariants of polynomial type (Alexander polynomial, Jones polynomial and their generalizations.).

6.3 Topological Quantum Field Theory - First Steps

In a very real sense, the connection of quantum mechanics with topology is an amplification of Dirac notation. In order to justify this idea of the amplification of notation, consider the following scenario. Let M be a 3-dimensional manifold. Suppose that F is a closed orientable surface inside M dividing M into two pieces M_1 and M_2. These pieces are 3-manifolds with boundary. They meet along the surface F. Now consider an amplitude

$$< M_1|M_2 >= Z(M).$$

The form of this amplitude generalizes our previous considerations, with the surface F constituting the distinction between the RpreparationS M_1 and the RdetectionS M_2. This generalization of the Dirac amplitude $< a|b >$ amplifies the notational distinction consisting in the vertical line of the bracket to a topological distinction in a space M. The amplitude $Z(M)$ will be said to be a topological amplitude for M if it is a toplogical invariant of the 3-manifold M. Note that a topological amplitude does not depend upon the choice of surface F that divides M.

From a physical point of view the independence of the topological amplitude on the particular surface that divides the 3-manifold is the most important property. An amplitude arises in the condition of one part of the distinction carved in the 3-manifold acting as Rthe observedS and the other part of the distinction acting as Rthe observerS. If the amplitude is to reflect physical (read topological) information about the underlying manifold, then it should not depend upon this particular decomposition into observer and observed. The same remarks apply to 4-manifolds and interface with ideas in relativity. We mention 3-manifolds because it is possible to describe many examples of topological amplitudes in three dimensions. The matter of 4-dimensional amplitudes is a topic of current research. The notion that an amplitude be independent of the distinction producing it is prior to topology. Topological invariance of the amplitude is a convenient and fundamental way to produce such independence.

7 Knot Logic

It is worthwhile to reflect on the structure that we have produced so far. We began with the topological problem of classifying knots and links in three-dimensional space. This problem is a direct abstraction of properties of closed loops of rope in the three-space of our experience, and the abstraction is designed to reflect directly back on those properties. That is, if we prove that the trefoil knot in our mathematical model is knotted, then this means that there can be no way to simplify a loop made in the form of a trefoil without tearing the loop. This is not a trivial remark. It is a remark of the same order as saying that a correctly performed *count* of the numbers from 1 to n will yield $n(n+1)/2$ for any chosen n. The mathematical model for integers touches our direct physical experience of counting. The mathematical model for knots touches our direct physical experience of topology.

As a result of this match with rope-work, the closed loop of rope under the manipulation of an observer becomes a precise analog computer for obtaining the solutions to certain concrete problems in knot theory. For example, the diagram shown in Figure 22 is unknotted. This can be determined by diagrammatic work with the Reidemeister moves, but it can also be determined at once, by threading a rope in this pattern and then pulling on the rope and finding that it unravels. Here the answer to the question "Is it knotted." is given but the steps in the process (in the sense of the Reidemeister moves) may be hidden in the motion of the rope. In establishing the diagrammatic formal system for knots and links we create a boundary between the world of physical observation of knots and this particular mathematical model. It is charming and a great boon to the intuition to move back and forth across this boundary.

One way to move across this boundary between the formal system corresponding to the knots and the underlying topology is to allow some flexibility in the interpretation of the diagrams that compose the formal system. For example, we can recognise that the diagram for a crossing indicates a (possibly) asymmetric relation between entities labelled by the overcrossing and the undercrossing arcs. The knot and link diagrams become indicators of the interweaving of these relations. Lets make this specific by calling the relation in question "membership" so that $a \epsilon b$ means that the arc a undercrosses the arc b. Then diagrams can indicate "sets" as shown in Figure 23.

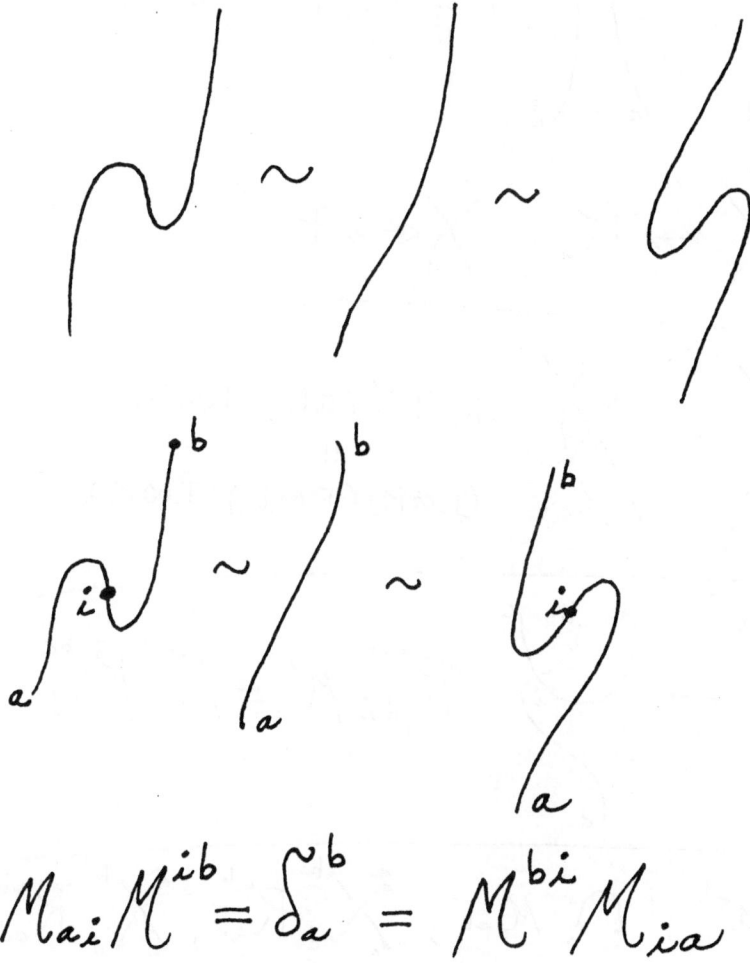

Fig. 20. Cancellation of Maxima and Minima

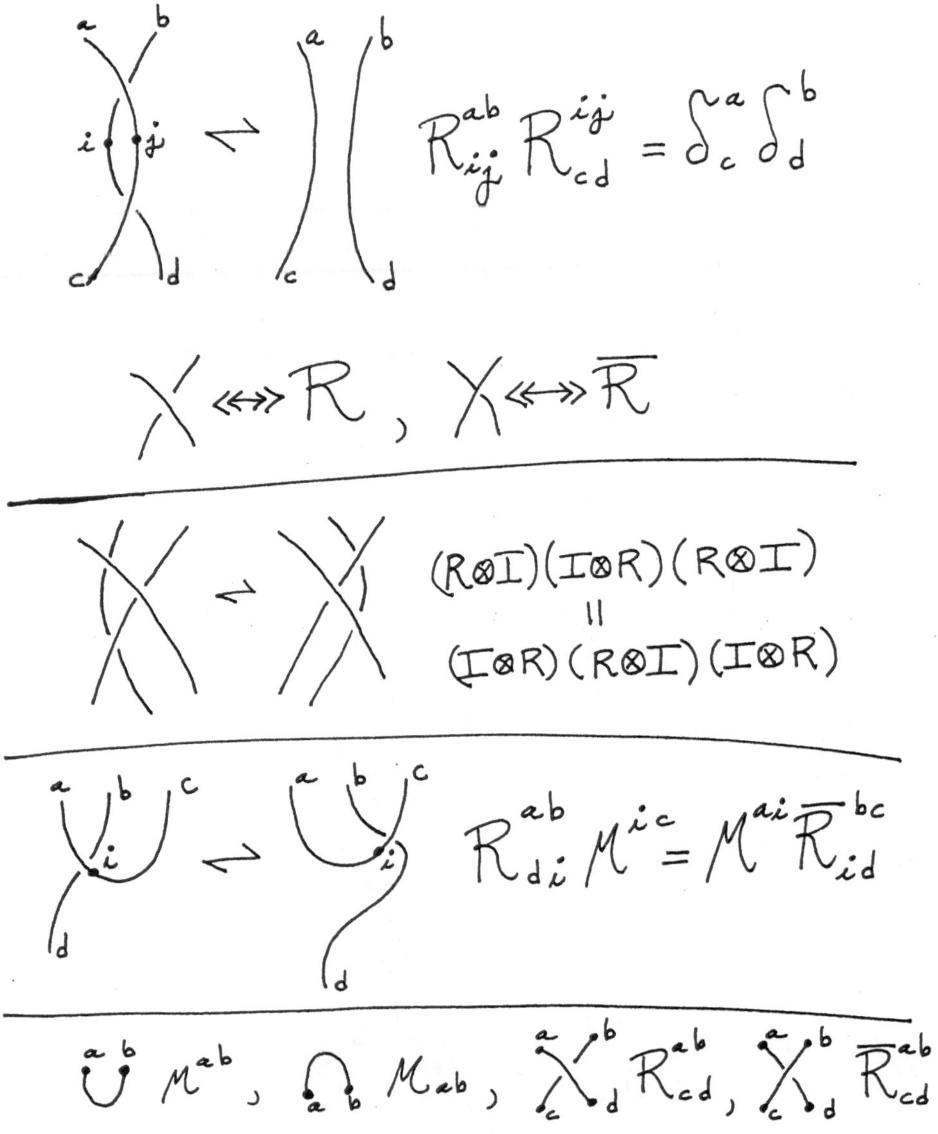

$$R^{ab}_{ij} R^{ij}_{cd} = \delta^a_c \delta^b_d$$

$$\times \Longleftrightarrow R \quad , \quad \times \Longleftrightarrow \overline{R}$$

$$(R \otimes I)(I \otimes R)(R \otimes I)$$
$$\|$$
$$(I \otimes R)(R \otimes I)(I \otimes R)$$

$$R^{ab}_{di} M^{ic} = M^{ai} \overline{R}^{bc}_{id}$$

$$\overset{a \ b}{\cup} M^{ab} \ , \quad \underset{a \ b}{\cap} M_{ab} \ , \quad \times^{a \ b}_{c \ d} R^{ab}_{cd} \ , \quad \times^{a \ b}_{c \ d} \overline{R}^{ab}_{cd}$$

Fig. 21.

Fig. 22.

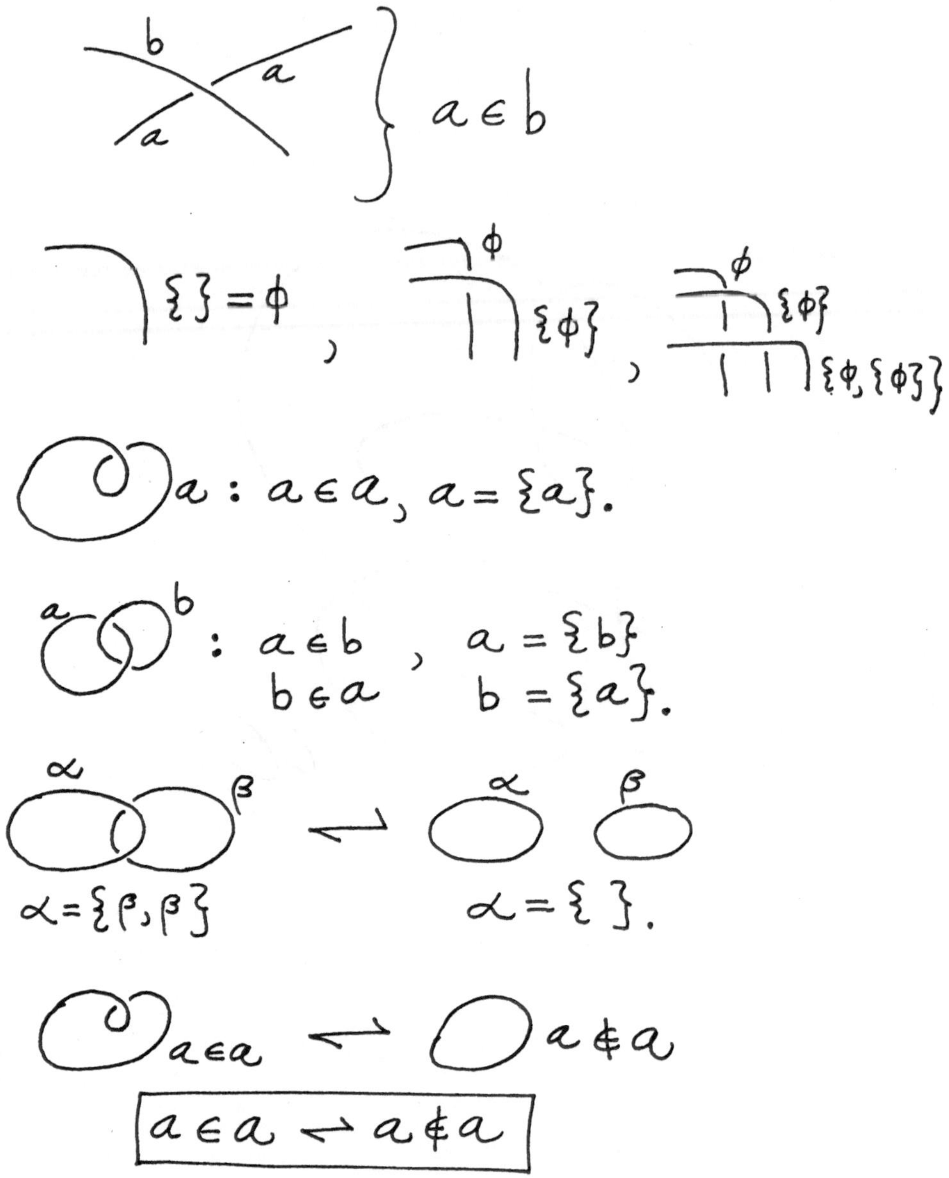

$$\left. \right\} \; a \in b$$

$$\bigg) \{ \} = \phi \quad , \qquad \bigg) \{ \phi \} \quad , \qquad \bigg) \{ \phi, \{ \phi \} \}$$

$$\bigcirc\!\!\!\!\bigcirc a : a \in a , \; a = \{ a \}.$$

$$\overset{a}{\bigcirc\!\!\!\bigcirc}{}^{b} : \; \begin{array}{l} a \in b \\ b \in a \end{array} , \quad \begin{array}{l} a = \{ b \} \\ b = \{ a \}. \end{array}$$

$$\overset{\alpha}{\bigcirc\!\!\!\bigcirc}{}^{\beta} \;\longleftharpoonup\; \overset{\alpha}{\bigcirc} \; \overset{\beta}{\bigcirc}$$

$$\alpha = \{ \beta, \beta \} \qquad\qquad \alpha = \{ \}.$$

$$\bigcirc\!\!\!\!\bigcirc \; a \in a \;\longleftharpoonup\; \bigcirc \; a \notin a$$

$$\boxed{ a \in a \;\longleftharpoonup\; a \notin a }$$

Fig. 23.

Note that these "knot sets" can be members of themselves (a curl), members of each other (a link), or simple hierarchies of membership such as the illustration in the Figure of diagrams for the von Neumann hierarchy that begins with the empty set, then the set consisting of the empty set, and continues with the next set consisting in the set of all the previously created sets. The knot sets give a model of set theory without the axiom of foundation. (The axiom of foundation states that a set cannot have infinite descending chains of membership.) Note however, that while a set that is a member of itself certainly has an infinite descent inherent in its structure, there is no need to regard this infinity as explicitly given in a model of the set. The model before us of the set whose member is itself as a "curl" is just such a model. The infinite descent corresponds to the fact that one can walk around and around a circle forever, but this path structure is a secondary property of the circle, not part of its primary structure. Just so, a computer program with a loop in it connotes a possible infinity, but the program itself is finite.

Since we began with topology, it is natural to ask whether the knot sets are related to the topology of knots and links. This requires a bit of thought. Clearly, we may not want to invoke the first Reidemeister move, since this will destroy self-membership. Note that we have already pointed out at the end of the previous section that the first Reidemeister move is special and can be regarded as encoding a framing of the knot or link. Nevertheless, I will take the first Reidemeister move as a move on the knot sets. This means that a set that is a member of itself is equivalent to as set that is not a member of itself. Specific representatives of a given set can be representatives of themselves if this is desired. The diagrammatic model shows us that self-membership is an *innocuous* issue, while mutuality such as $a = \{b\}$ and $b = \{a\}$ is a more serious structure directly tied to the image of linking. Mutuality will not vanish just because we allow flexibility in self-reference. With the rule that

$$a = \{a, b, c, ...\} \Longleftrightarrow a = \{b, c, ...\}$$

we have that a set is a member of itself if and only if it is not a member of itself (!) and Russell's paradox cannot arise.

As for the third Reidemeister move, it does not affect the relations in the corresponding knot set. Hence knot sets are invariant under the third Reidemeister move.

But what about the second Reidemeister move? Consider the diagram in Figure 25. In this diagram we see that the multiset $\{b, b\}$ is transformed into the null set when the second Reidemeister move is performed on the diagram.

Two issues are brought forward by this example. We see that a diagram may represent a set with a multiplicity of identical elements (a multi-set). In standard set theory it is common practice to reduce a multi-set to a set by *condensing* all the occurrences of an entity to a single representative of that entity. Thus $\{b, b\}$

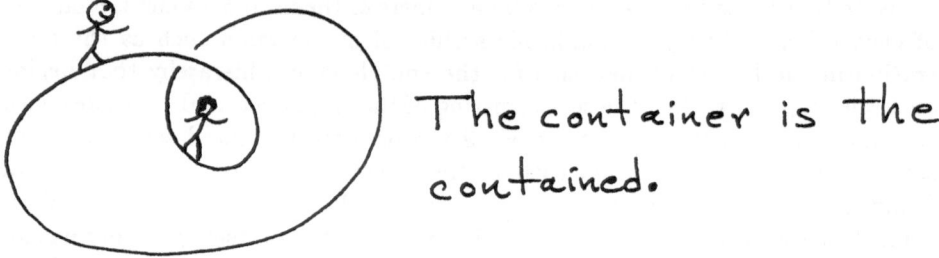

Fig. 24. The container is the contained.

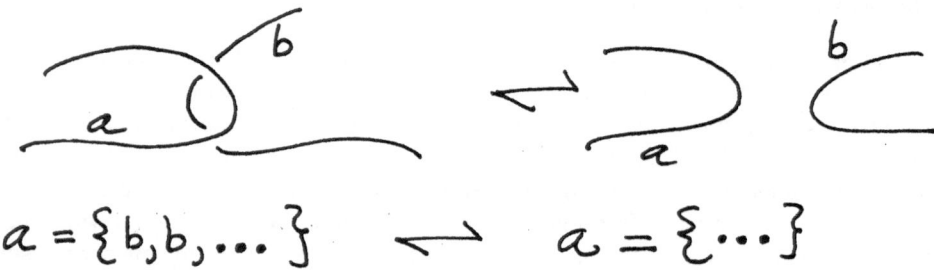

Fig. 25. Identicals cancel in Pairs

is equivalent to $\{b\}$ in standard set theory. However, we have spoken with the second Reidemeister move, and she has told us to adopt a different rule .

Rule of Cancellation of Identicals: *Identicals cancel in pairs.*

With this rule in place, $\{b, b\}$ is equivalent to $\{\}$, and the knot sets are invariant under the second Reidemeister move, and hence they are invariants of ambient isotopy of link diagrams. Throw a knot set into the air, catch it and place it back onto the plane. All relations of membership are preserved in this topological transition.

8 A Knot-Logical Dialogue Between Inquisitive Alpha and her Knot Theorist Friend Beta

Alpha: So you think that you have solved the Russell paradox with this knot set theory? How is your solution any different than banning speech about self reference?

Beta: Mine is more realistic in a psychological sense. A knot set S is an entity that is capable of two mutually exclusive interpretations.

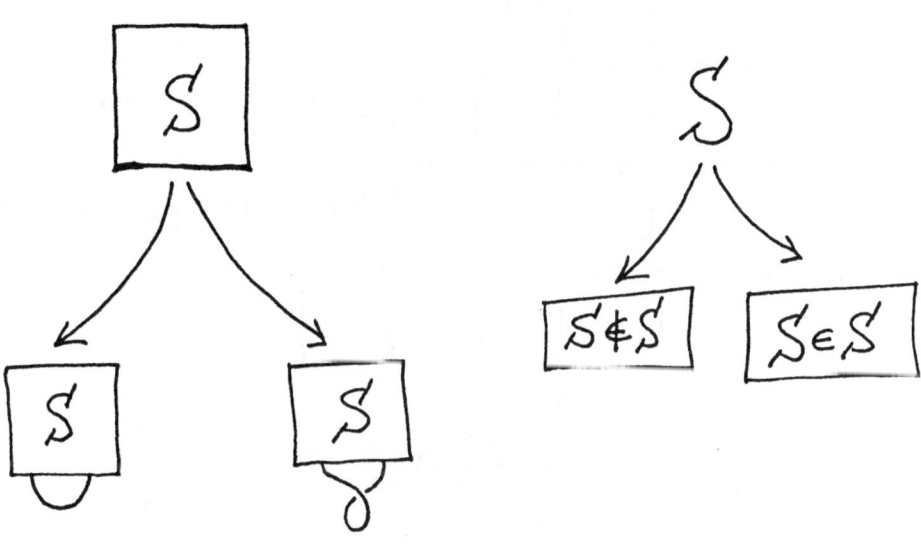

Fig. 26.

These two interpretations are two ways to observe the set S.

Alpha: Then you are saying that $a \epsilon a$ is not a *property* of a, but rather it is only a property of some way of observing a?

Beta: Yes. Observation is the same as projection of the "actual" link in 3-space to the planar diagram. Mutuality of membership occurs in every projection of the simple link of two circles. I am willing to say that P is a *property* of a knot set S if P is true for every projection of S. I will further say that P is a *possibility* for S if P is true for some projection of S. Thus every set has the possibility of self-membership, but self-membership is not a property of the set.

Alpha: I see what you mean, but it does seem that you are restricting yourself to finite knot sets. What if I were to consider an infinite limit of the von Neuman construction in the form of a knot set (*See Figure 27*). Then this limit has the form $S = \{...SSS\}$, a multi-set containing an infinite number of copies of itself! You will not be able to get this set to refuse self-membership.

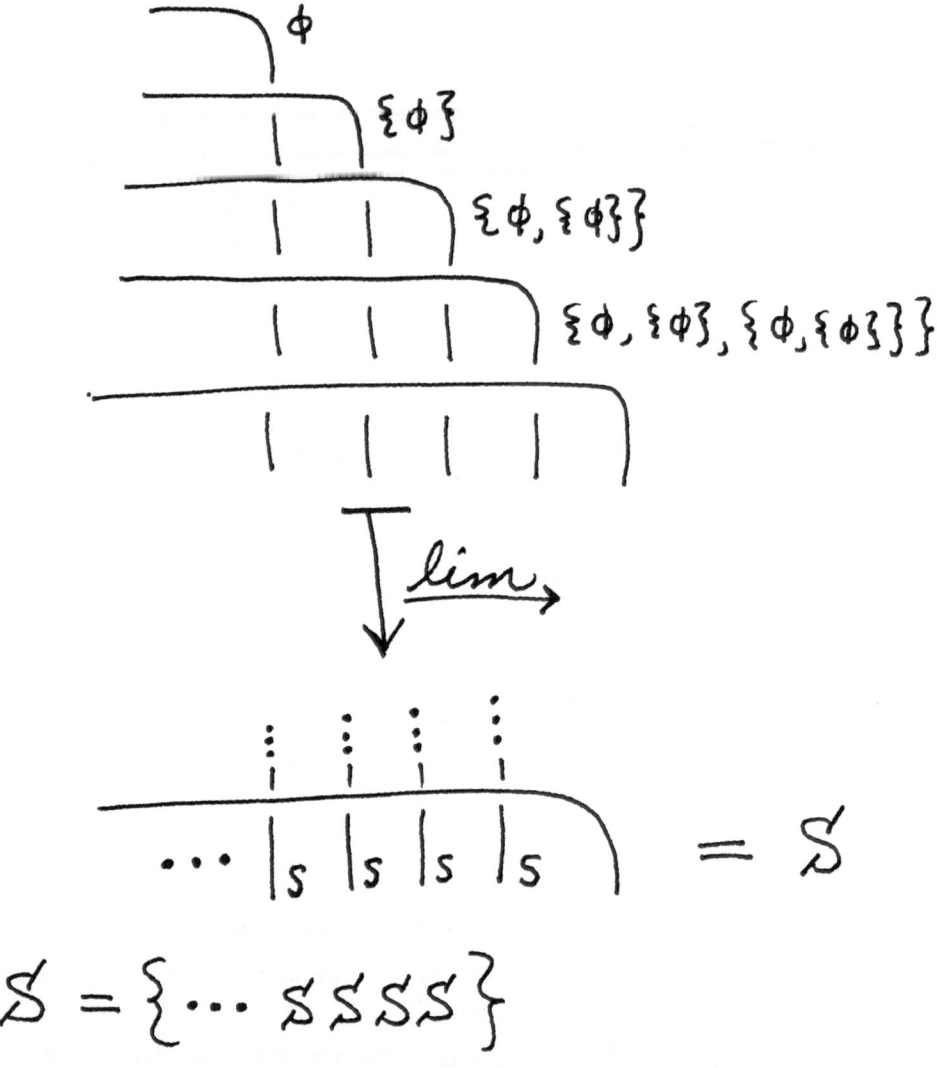

Fig. 27. The von Neumann Limit

Beta: You will just have to allow me to make pairwise cancellation infinitely often! Then

$$...SSSS = ...(SS)(SS) = \text{``nothing''}$$

while
$$...SSSS = ...SSSSS = ...(SS)(SS)S = \text{"nothing"}S = S.$$

Thus $S = \{...SSSS\} = \{\}$ and $S = \{...SSSSS\} = \{S\}$, just in accord with the topological philosophy of knot-sets. We are dancing on the boundary of logic.

Alpha: This makes my head spin.

Beta: But we have just begun! Lets go back to knot theory proper and I will show you how this "infinite repetition trick" can be used to deduce real and tangible properties of knots. Consider the question:

Can two knots cancel each other?

That is, if I tie one knot on the rope and then tie the second knot on the rope immediately after the first knot. Can these two knots topologically undo each other? (As usual I will regard the two knots as living on a closed loop of rope in three space.)

Alpha: I don't know. This sounds like one of your trick questions. I have seen magicians make knots disppear by making further weaves in a given knot.

Beta: No. I do not allow that magician's trick business. The two knots are fully separate from one another. First you tie one and then you tie the other one. Like this.

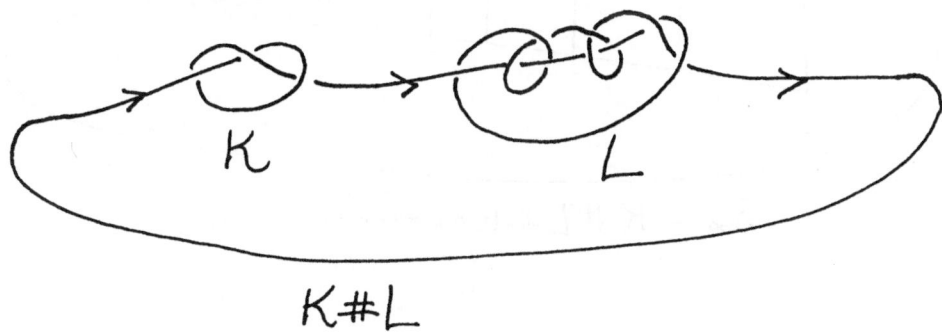

Fig. 28.

Alpha: Well, then I do not think that two knots can cancel each other.

Beta: You are right! Can you prove it? Would you like to see a proof?

Alpha: [After some thought.] Well all right, let me see your proof. [muttering to herself] Proofs, proofs, why should I believe these proofs.

Beta: Here is the proof. Suppose that the two knots are denoted by K and L. Let $K\#L$ denote the result of first tieing K and then tieing L. This is called the connected sum of K and L. Now form the *infinite* connected sum of K and L which I shall denoted by S_∞:

$$S_\infty = K\#L\#K\#L\#K\#L\#K\#\cdots$$

I make this infinite knot by making the successive copies of K and L smaller and smaller so that they converge to a limit point P in three-space. Then from this limit point P I draw a curve that returns to the starting point of the left-most K. The result is a "wildly" embedded circle in three-space with the desired infinite knotting in it.

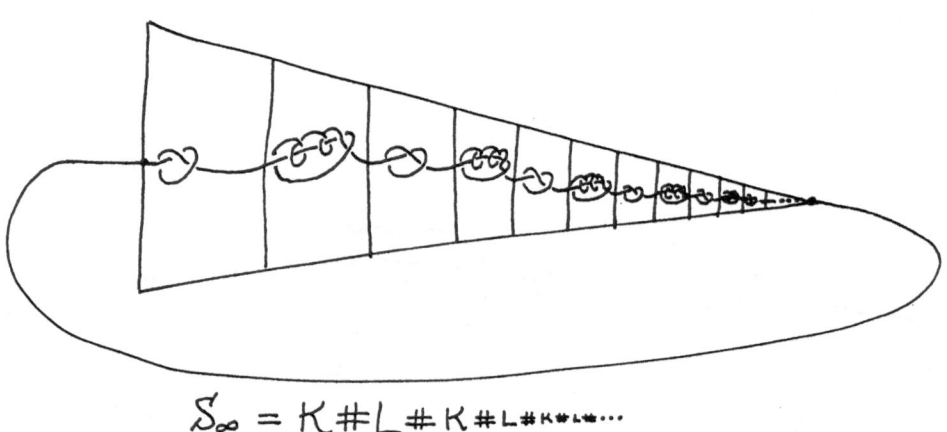

$$S_\infty = K\#L\#K\#L\#K\#L\#\cdots$$

Fig. 29.

Now suppose that the original knots K and L can cancel one another. Then we have

$$K\#L = U$$

where U is a trivial knot so that

$$U\#A = A\#U = A$$

for any knot A, tame or wild. Incidentally, you easily check that $K \# L = L \# K$ for any K and L by making K small and threading it through L until it ends up on the right of L. Thus if $K \# L = U$ then $L \# K = U$ as well. Then we have

$$S_\infty = (K \# L) \# (K \# L) \# (K \# L) \# ... = U \# U \# U \# ... = U$$

and

$$S_\infty = K \# (L \# K) \# (L \# K) \# (L \# K) \# ... = K \# U \# U \# U \# ... = K$$

Thus

$$K = S_\infty = U.$$

Similarly,

$$L = U.$$

So the only way that two knots can cancel each other is if they are each unknotted to begin with. How do you like this real-life application of our considerations about the problems of self-membership in infinite knot sets?!

Alpha: I like the form of this proof, but aren't you doing something sneaky by using the wild knot. It seems to me that you proved that $K = U$ and $L = U$ only by going through the category of wild knots. How do you know that K and L might not be really knotted as tame knots but somehow can be undone in your "wild" category?

Beta: Precisely the right question dear Alpha. In fact there is a theorem that tells us that two tame knots that are isotopic through the wild category are in fact tamely isotopic [17]. I cannot do the proof of *that* theorem here, but thinking about it will give you food for thought. The possibility of using infinite constructions with logical validity is one of the things that gives topology its zing.

Alpha: Your wild knot , $S_\infty = K \# L \# K \# L \# K \# L \# K \# ...$, is a fixed point of the function on knots defined by the equation

$$F(X) = K \# L \# X.$$

Thus

$$F(S_\infty) = S_\infty.$$

This reminds me of a logical construction that builds fixed points: You start with a function ,F, and define a new function G by the formula

$$GX = F(XX).$$

Then

$$GG = F(GG).$$

Hence any function has a fixed point![2]! This is called the basic fixed point theorem of the untyped lambda calculus and it is the foundation of recursion theory. Of course, it is not far from saying that the fixed point of F is

$$F_\infty = F(F(F(F(F(F(...)))))),$$

the infinite composition of F with itself. The three dots in the infinite composition mean "more of the same" and the *syntax* of the repetition XX creates the same effect without the disturbing idea of writing a symbol with infinitely many parts. Each of these fixed points (GG and F_∞) is beautifully written and content-free.

Beta: Wait a minute! What do you mean by XX? If X is a knot and XX is another knot, then what do you mean by GG? After all, G is not a knot. Anyway, my fixed point S_∞ is not content-free. It is a knot with an infinte weaving pattern.

Alpha: Well, your S_∞ is not really a knot either. It is certainly not a knot that anyone could actually tie in a rope. The fixed point theorem in the lambda calculus is actually a very concrete and down-to earth business. Let me borrow the idea of diagrams from you and illustrate it. Let G denote the following glyph.

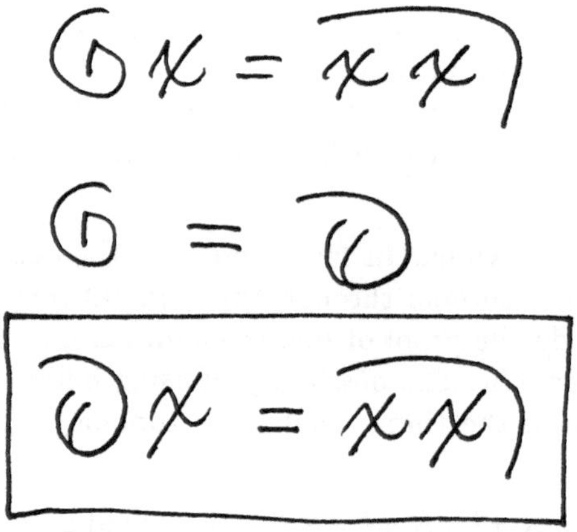

$$\text{G} X = \overline{X X)}$$

$$\text{G} = \text{(0)}$$

$$\boxed{\text{(0)} X = \overline{X X)}}$$

Fig. 30.

G acts on another glyph X by *making two copies of X and placing them next to one another underneath a little roof.*

Fig. 31.

But now you see that if we let G act on G, then G will make two copies of itself and place them underneath the little roof. Thus G gets *replaced* by *two* copies of G underneath a little roof, and we write

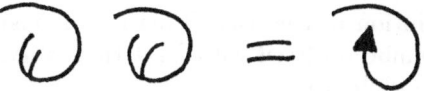

Fig. 32.

Beta: Aha! When you write the equals sign, you mean "is replaced with." You do not mean "equals." I have no problem with a rule that says that GG is replaced with $F(GG)$. That does not mean that GG and $F(GG)$ are actually the same. But my knot S_∞ and the knot $K\#L\#S_\infty$ are actually the same knot.

Alpha: Lambda calculus is above all that. You have to do a mighty contortion of your language to assert that those "knots" are "the same." You have to admit infinite knotting as actually real. I think that this is a mathematician's fantasy. In ordinary mathematical reality, I can always tie $K\#L$ one more time, but this does not mean that S_∞ actually exists. In my view S_∞ is just a way of saying that we can weave the knot one more time. Why I would not even say that $X = X$ means that X is identical with itself. It just means that X can be replaced by a simulacrum of itself.

Beta: Well I would say that a knot undoubtedly is itself and the universe is also itself. But in order to be seen, the universe has to be divided into a part that sees and a part that is seen. In the case of the knot and our perception of it, this means that the knot is broken up into the spectrum of parts that is its projection in the plane. That projection, like the shadows on the walls of Plato's cave is not the knot, but it is a powerful container of information about the knot. In the same way, my infinite knot with its projection with an infinite number of parts is but a single weaving pattern in three dimensional space. I regard it as a unity that happens to decompose into these parts. Mathematics does not come fully alive for anyone until they see that infinities are not just processes of replacement. Infinities are actual unities. They are whole forms that really exist!

Alpha: That was spoken like a true geometer. And I certainly sympathize with your sentiments, but I have been assigned the role of the skeptic and I have to ask you to note carefully that your infinite actualities are constructed with great analogical care to behave as best they can as though they had the properties of ordinary objects (In my view there are no ordinary objects.). I have to deal with Russell and his paradox. Suppose that X and Y are sets and that XY means that "X is a member of Y" while $\sim X$ means *not X*. Now define RX by the equation

$$RX =\sim XX.$$

As you can see, R defines a set whose members are exactly those sets that are not members of themselves. Substituting R for X, I obtain a fixed point of negation:

$$RR =\sim RR$$

But for me there is no problem with this. It just says that I can replace the negation of RR with RR. It does not mean that they are the same! In fact, you yourself have explained to me that your solution to the Russell paradox is just this, that you can replace a set that is a member of itself with a set that is not a member of itself without running into any difficulty.

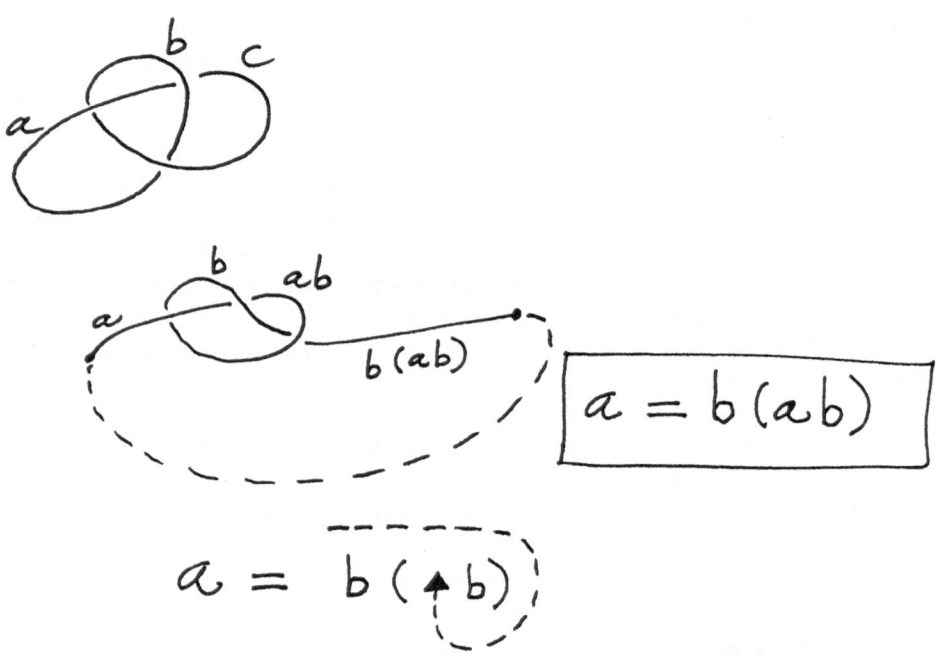

Fig. 33.

Beta: Yes, but my solution goes outside the cave of shadows and shows that the whole idea of self-membership can be regarded as an illusion of the projection. In fact, your lambda calculus and my knots are really very close relatives. You will recall all those lectures that I gave you about the IQ (the involutory quandle - see section 2) of a knot. The IQ of a knot is a non-associative algebra, just like your lambda algebra, and in fact each knot can be seen as a fixed point in this algebra. For example, take the trefoil with generators a,b,c and relations $a = bc$, $b = ca$, $c = ab$. We can write $a = b(ab)$ and regard this as a description of the knot via a as a fixed point for the operator $FX = b(Xb)$. In fact each arc of the knot is a fixed point and through this fixed point describes the knot from its own point of view. It is clear to me that the knot diagrams provide an excellent place for depicting the intricacies of the lambda calculus!

Alpha: You are beginning to convert me. We had better take a break before I lose my skepticism.

9 Coda

This essay has concentrated on the intricate relationship between invariants of knots, quantum amplitudes and the logic of paradox and infinity, relations that abound in the study of complex systems of all kinds. Knots provide an extraordinary ground in the study of complex systems exactly because they partake of natural geometry,topology and an internal mathematical structure of great depth and beauty.

References

1. P. Aczel, *The Theory of Non-Well-Founded Sets*, 1988,CLSI Lecture Notes, No. 14.
2. H. P. Barendregt, *The Lambda Calculus Its Syntax and Semantics*, North Holland,1981,1985.
3. "From large cardinals to braids via left distributive algebra", (to appear in the Journal of Knot Theory and Its Ramifications).
4. R.A. Fenn and C.P. Rourke, "Racks and links in codimension two", *J. Knot Theory and its Ramif.*, 1992, Vol.1, No.4, pp. 343-406.
5. Frederic B. Fitch, *Elements of Combinatory Logic*, New Haven and London, Yale University Press,1974.
6. R. H. Fox, *Introduction to Knot Theory*, Blaisdell Pub. Co., 1963.
7. V.F.R. Jones, "A polynomial invariant of links via von Neumann algebras", *Bull. Amer. Math. Soc.*, 1985, No. 129, pp. 103-112.
8. D.Joyce, "A classifying invariant of knots,the knot quandle", *J. Pure and Appl. Algebra*, 1983, Vol. 23, pp. 37-65.
9. L.H. Kauffman, "State models and the Jones polynomial", *Topology*, 1987, Vol. 26, pp. 395-407.
10. L. H. Kauffman, "Self-Reference and Recursive Forms", *Journal of Social and Biological Structures*, 1987,vol.10, pp. 53-72.
11. L. H. Kauffman, *Knots and Physics*, World Scientific Pub., 1991,1994.
12. L. H. Kauffman and S. L. Lins, *Temperley-Lieb Recoupling Theory and Invariants of 3-Manifolds*, Annals of Mathematics Study 114, Princeton Univ. Press,1994.
13. L. H. Kauffman, "Knot Logic", To appear in *Knots and Applications*, edited by L. Kauffman, World Scientific Pub., 1994.

14. L.H. Kauffman and S.W. Winker, *Quandles, Crystals and Racks - A New Approach to Knot Theory*, (book in preparation), World Scientific Pub.

15. "A calculus for framed links in S³", *Invent. Math.* 45(1978),pp. 35-56.

16. W.B.R.Lickorish, "A representation of orientable, combinatorial three-manifolds", *Ann. of Math.* **76** (1962), pp. 531-540.

17. E.E. Moise, *Geometric Topology in Dimensions Two and Three*, Springer Verlag, New York, 1977.

18. A. Pedretti, *Self-Reference on the Isle of Wight - Transcripts of the First International Conference on Self-Reference*, (August 24-27, 1979), Princelet Editions London and Zurich.

19. J. H. White, "Self-linking and the Gauss integral in higher dimensions", *Amer. J. Math.* **91** (1969), pp. 693-728.

20. S.W. Winker, *Quandles, Knot Invariants and the n-fold Branched Cover*, (1984), Doctoral Thesis, Univ. of Illinois at Chicago.

21. Edward Witten, "Quantum field theory and the Jones Polynomial", *Commun. Math. Phys.*,vol. 121, 1989, pp. 351-399.

Part IV
Towards a Theory of Landscapes

Towards a Theory of Landscapes

Peter F. Stadler

Institut für Theoretische Chemie, Universität Wien.
Währingerstraße 17, A-1090 Wien, Austria

Santa Fe Institute.
1399 Hyde Park Rd., Santa Fe, NM 87501, USA

1 Introduction

Since Sewall Wright's seminal paper [1] the notion of a *fitness landscape* underlying the dynamics of evolutionary adaptation optimization has proved to be one of the most powerful concepts in evolutionary theory. Implicit in this idea is a collection of genotypes arranged in an abstract metric space, with each genotype next to those other genotypes which can be reached by a single mutation, as well as a value assigned to each genotype. Such a construction is by no means restricted to biological evolution; Hamiltonians of disordered systems, such as spin glasses [2, 3], and the cost functions of combinatorial optimization problems [4] have the same mathematical structure. It has been known since Eigen's [5] pioneering work on the molecular quasispecies that the dynamics of optimization on a landscape depends crucially on detailed structure of the landscape itself. Extensive computer simulations, see, e.g., [6, 7, 8], have made it very clear that a complete understanding of the dynamics is impossible without a thorough investigation of the underlying landscape [9].

The landscapes of a number of well known combinatorial optimization problems[1] such as the Traveling Salesman Problem TSP, the Graph Bipartitioning Problem GBP, or the Graph Matching Problem GMP have been investigated in some detail [10, 11, 12]. The distribution of local optima and the statistical characteristics of down-hill walks have been computed for the uncorrelated landscape of the random energy model [13, 14, 15, 16]. Furthermore, two one-parameter families of tunably rugged landscapes have been studied extensively: the Nk model and its variants [17, 18, 19, 20] and the p-spin models [8, 21]. A detailed survey of a variety of model landscapes derived from folding RNA molecules into their secondary structures has been performed recently [22, 6, 7, 23, 20, 24, 25, 26, 27, 28, 29, 30, 31, 32, 33].

In this contribution I will present an overview over the mathematical techniques that have proved useful for an analysis of landscapes. This approach is based on Fourier Series defined on the highly symmetric graphs that occur as

---

[1] See section 4.4 for details.

configuration spaces underlying combinatorial optimization problems and evolutionary dynamics. We will consider both stochastic models of landscapes, so-called random fields, and individual "measured" landscapes.

Section 2 deals with the structure of the configuration spaces. The mathematical language described there, mostly permutation group theory and graph theory, seems to be the appropriate basis for a general theory of (fitness) landscapes.

Section 3 discusses random fields. The theory of random fields at this points is mostly a "second order theory". That is to say, it revolves around covariance and correlation measures. The main emphasis lies here on *isotropic* random fields and their correlation functions. Among other results we give a characterization of isotropic random fields in terms of their Fourier coefficients. Then we briefly discuss the relation between the autocorrelation function of random fields and the correlation function of the time series obtained from a random walk on the random field. Furthermore we investigate superpositions and transformations of random fields.

Section 4 contains the theory of individual landscapes. The most surprising result concern the relations between "elementary landscapes",which are eigenvectors of the graph Laplacian Δ, and their autocorrelation functions. The relation of the correlation structure and the distribution of local optima of the landscape is subject of on-going research; first results are discussed in subsection 4.3. The notion of elementary landscapes leads to a refined classification scheme for landscapes. The final subsections contain material about landscapes on irregular graphs and on anisotropies in landscapes on highly regular graphs.

Section 5 provides a condensed review of RNA landscapes and a generalization of the theory of landscapes to sequence-structure mappings.

2 Configuration Spaces

2.1 An Example — TSP

The travelling salesman problem (TSP) [34] is a classical example of an NP-complete [4] combinatorial optimization problem. Few mathematical problems have attracted as much attention. Given a distribution of cities the task is to find the shortest tour visiting each city once and returning to the starting point with prescribed costs w_{ij} for traveling from i to j. The cost function is

$$f(\tau) = \sum_{i=1}^{n-1} w_{\tau(i)\tau(i+1)} + w_{\tau(n)\tau(1)}$$

where τ is the permutation encoding the order of the cities. The symmetric problem $w_{ij} = w_{ji}$ has applications in X-ray crystallography [35], electronics [36], and the study of protein conformations [37]. The asymmetric case [38] has applications in scheduling chemical processes or from pattern allocation problems in the glass industry.

Besides the cost function there is another ingredient in the problem that is as important: the combinatorial structure of the configurations (in our case the possible tours of the salesman). Heuristic algorithms designed to find approximate solutions of the TSP use the internal structure of configurations to arrange them (hopefully) in a way which simplifies the search. Typically the search proceeds from configuration to a neighboring one.

What does neighboring mean in the case of the TSP? There is no unique answer, instead a few different ways of defining neighborhood come to mind. For a mathematician the most natural choice is probably to use the exchange of two cities along the tour (a transposition) as elementary "move". Alternatively one might allow only the exchange of subsequent cities; this operation is called a "canonical transposition". Lin and Kernighan [39] suggested to reverse the order of the cities in a part of the tour, this move is called inversion, reversal, or 2opt move. These three types of moves are displayed in figure 1.

All landscapes are composed of these two ingredients:

1. a *cost function* or *fitness function* f assigning a value to each *configuration*, and
2. a rule determining whether two configurations x and y are neighbors of each other.

As a consequence of 2, the set C of all configurations can be viewed as a graph Γ with each vertex corresponding to a configuration and an edge between any two nearest neighbors. A landscape is therefore simply a real valued function of the vertex set V of some graph Γ. A list of corresponding terms for landscapes occurring in different fields is presented in table1.

Table 1. The Correspondence of Statistical Mechanics, Combinatorial Optimization and Evolutionary adaptation.

Combinatorial Optimization	Statistical Mechanics	Folding of Biopolymers	Evolutionary Adaptation	Symbol
system size	system size	chain length	chain length	n
configuration	micro-state	structure	sequence	x
cost	energy	energy	fitness	$f(x)$
pseudo-temperature	temperature	temperature	mutation-rate	$T = -\ln q$
optimum	ground state	native state	fittest	o
instance	sample	*	*	f
move	(spin flip)†	†	mutation	

† There is no generic term for the elementary process of the dynamics.
* Ideally, there is only one energy or fitness function in a given system.

Often we are not given a single landscape but a model containing a large number of parameters which are assigned randomly. It is customary, for instance,

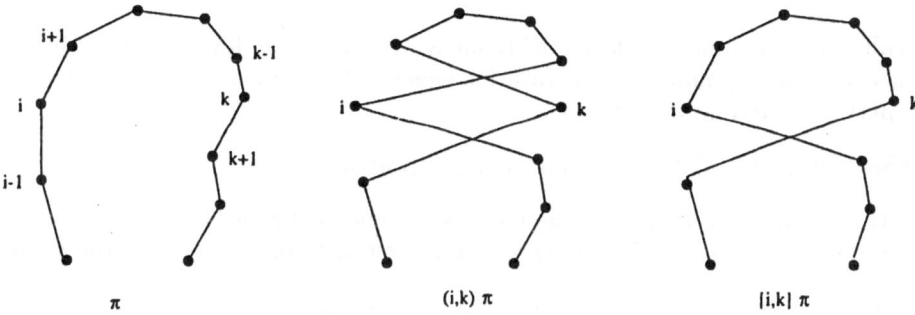

Fig. 1. The two most common types of elementary moves for the TSP are transpositions, $(i, k)\pi$, and inversions, $[i, k]\pi$ of a tour π. A transposition of the form (i,i+1) is called *canonical transposition*.

to consider TSPs with the cities randomly distributed in the unit square [34]. Such a model is not a landscape but consists of an entire probability space whose elements are landscapes. These so-called *random fields* will be discussed in section 3.

2.2 Groups

Permutation Groups. In this section we briefly recall the basic definitions of the theory of permutation groups as far as we will need them in this contribution. An excellent introduction to permutation groups is Wielandt's book [40]. Let X be a finite set and G a group acting on X, notation (G, X). The group identity will be denoted by e. The cardinality of X is the *degree N* of the permutation group. The cardinality of G itself is called the *order* of G. The *permutation matrix* associated with a group element $g \in G$ is

$$g_{ij} \stackrel{\text{def}}{=} \begin{cases} 1 & \text{if } g(i) = j \\ 0 & \text{otherwise} \end{cases}$$

In this notation we identify an element of the abstract group G with its representation as a permutation matrix. Analogously, we will write G also for the group of permutation matrices corresponding to the group elements $g \in G$.

$D(x) \stackrel{\text{def}}{=} \{y \in X \,|\, y = g(x), \forall g \in G\}$ is an *orbit* of G in X. The orbits of G partition the set X. We have $y \in D(x) \implies D(y) = D(x)$.

The *centralizer* of G in the algebra of all $N \times N$ matrices over $\mathbb{C}$, i.e.,

$$\mathfrak{C}[G, X] \stackrel{\text{def}}{=} \{Q \in \mathbb{C}^{N \times N} \,|\, Qg = gQ \quad \text{for all} \quad g \in G\},$$

is called the *commuting algebra* of G on X. It seems to play a key role in the study of highly symmetric configuration spaces. We will also need a few simple properties of permutation groups.

Definition 1. Let G be a permutation group acting on X.

- G acts *transitively* in X if there is only a single orbit of G.
- G acts *regularly* on X if $g(x) = x$ for some x implies that g is the group identity.
- A subset $Y \subset X$ such that for each $g \in G$ we have either $g(Y) = Y$ or $g(Y) \cap Y = \emptyset$ is called a block. The one-point subsets of X, and X itself, clearly fulfill this requirement for all permutation groups; they are usually referred to as trivial blocks. G acts *primitively* on X if there are no non-trivial blocks.

Orbits, Orbitals, and Symmetry Classes. If G acts transitively on X it does not necessarily do so[2] on $X \times X$. We will denote the orbits of G on $X \times X$ by $\ddot{\kappa}$, $\ddot{\mu}$, $\ddot{\nu}$, etc.

There is a one-to-one relationship between the orbits of G on $X \times X$ and the orbits of the *stabilizer subgroup*

$$G_0 \stackrel{\text{def}}{=} \{g \in G \,|\, g(0) = 0\}$$

of an arbitrary element $0 \in X$:

$$\dot{\mu} = \{v \in X \,|\, (v, 0) \in \ddot{\mu}\},$$
$$\ddot{\mu} = \{(u, v) \in X \times X \,|\, \exists \alpha \in G : \alpha(v) \in \dot{\mu} \wedge \alpha(v) = 0\}.$$

An *orbital* is a mapping Ψ from X into the subsets of X such that

---

[2] $g(x, y) \stackrel{\text{def}}{=} (g(x), g(y))$

(i) $\Psi(x)$ is a G_x-orbit $x \in X$.

(ii) $g(\Psi(x)) = \Psi(g(x))$.

Let $\Psi_\mu(x)$ denote the orbitals, i.e., $\Psi_\mu(x)$ is the orbit of the stabilizer G_x of x that corresponds to the orbit μ of G_0, the stabilizer of the "reference vertex" 0. We have therefore $x \in \Psi_\mu(y)$ if and only if $(x, y) \in \ddot{\mu}$. We will refer to μ as the *symmetry class* to which the orbits μ and $\ddot{\mu}$ as well as the orbital Ψ_μ belong. The mirror image of an orbital is defined by

$$\Psi^+(x) = \{g^{-1}(x)|g(x) \in \Psi(x)\},$$

it is again an orbital of G acting on X. Consequently, for each μ there is a ν such that $\Psi_\mu^+ = \Psi_\nu$. We write $\nu = \mu^+$.

The symmetry class corresponding to the diagonal on $X \times X$ will be denoted by 0, i.e., $\ddot{0} = \{(x, x)|x \in X\}$. The number of symmetry classes is $M + 1$.

The cardinalities of the sets $\Psi_\mu(x)$ are of course independent of x. It is often called a *subdegree* of G; it will be denoted here by $|\mu|$.

Incidence Matrices. A central part of the theory discussed in this contribution are the matrices $R^{(\mu)}$ defined by

$$R_{xy}^{(\mu)} = \begin{cases} 1 & \text{if } (x, y) \in \ddot{\mu} \\ 0 & \text{otherwise} \end{cases} = \begin{cases} 1 & \text{if } x \in \Psi_\mu(y) \\ 0 & \text{otherwise} \end{cases}$$

Higman [41] called them *incidence matrices* of the transitive permutation group G on X. The mirror image of a symmetry class belongs to the transpose of the corresponding incidence matrix: $(R^{(\mu)})^+ = R^{(\mu^+)}$. The collection of the $M + 1$ matrices $R^{(\mu)}$ will be denoted by $\mathfrak{R}$.

Definition 2. Let $\mathfrak{A} = \{A_0, A_1, \ldots, A_M\}$ be a collection of $N \times N$ 0-1 matrices with the following properties:

(i) $A_0 = I$, the identity matrix

(ii) $\displaystyle\sum_{j=0}^{M} A_j = J$, the matrix with all entries 1.

(iii) For all $0 \le i, j \le M$ we have $A_i A_j = \displaystyle\sum_{k=0}^{M} s_{ijk} A_k$.

(iv) For all $0 \le i \le M$ there is a j such that $A_i^+ = A_j$.

Then we call $\mathfrak{A}$ an *incidence scheme*.

Properties (i), (ii), and (iii) imply that $\mathfrak{A}$ forms the basis of an algebra of dimension $M + 1$; we will denote this algebra by $\langle \mathfrak{A} \rangle$.

A well known result is the following:

Theorem 3. *Let X be a finite set and let G be a permutation group operating transitively on X. Then the collection of the matrices $\mathfrak{R}$ forms an incidence scheme. In fact, the collection $\mathfrak{R}$ generates the commuting algebra of G on X:*

$$\langle \mathfrak{R} \rangle = \mathfrak{C}[G, X].$$

Proof. It is trivial that (i), (ii), and (iv) are satisfied. It is shown in [40, Thm. 28.4] and [42, Thm. 4.3.6] that the matrices $R^{(\mu)}$ form a basis of the commuting algebra $\mathfrak{C}[G, X]$, see also [41], and thus (iii) holds as well. $\square$

Intersection Numbers and Collapsed Matrices. We define the *intersection numbers* of the orbitals Ψ_κ and Ψ_μ with respect to the orbital Ψ_ν by

$$s_{\kappa\nu}^{(\mu)} \stackrel{\text{def}}{=} |\Psi_\mu(u) \cap \Psi_\kappa(w)| \qquad \text{with} \quad u \in \Psi_\nu(w).$$

It is easy to see that these numbers really depend only on κ, μ, and ν [41]. The *intersection matrix* of the orbital Ψ_μ is the matrix with entries $(s_{\kappa\nu}^{(\mu)})$.

The intersection numbers are of course closely related to the incidence matrices. In fact, we have the following

Lemma 4. $R^{(\mu^+)} R^{(\nu)} = \sum_\kappa s_{\nu\kappa}^{(\mu)} R^{(\kappa)}$.

Proof. We have $[R^{(\mu^+)} R^{(\nu)}]_{xy} = \sum_z R_{xz}^{(\mu^+)} R_{zy}^{(\nu)} = \sum_z R_{zx}^{(\mu)} R_{zy}^{(\nu)}$

$= \left| \{ z | z \in \Psi_\mu(x) \wedge z \in \Psi_\nu(y) \} \right| = |\Psi_\mu(x) \cap \Psi_\nu(y)| = s_{\nu\kappa}^{(\mu)},$
where κ is such that $x \in \Psi_\kappa(y)$, i.e., it is the coefficient of R_{xy}^κ. $\square$

The intersection matrices provide an elegant means for determining if G acts primitively on X:

Lemma 5. *[41, (4.8)]* G *acts primitively on X if and only if all intersection matrices $R^{(\mu)}$ are irreducible.*

Higman [41] shows that the intersection matrix $S^{(\mu)}$ can be obtained from the incidence matrix $R^{(\mu)}$ by the following procedure:
Suppose the points of X are arranged according to the symmetry classes μ, and consider the corresponding blocks of $R^{(\mu)}$ (see the example below). Each block has by construction constant column sum. Construct the matrix $\hat{R}^{(\mu)}$ by replacing each block by its column sum. Using the terminology in [43] we call a matrix obtained by this procedure *collapsed*.

Theorem i *The collapsed incidence matrices are the intersection matrices, i.e.,*
$\hat{R}_{\kappa\nu}^{(\mu)} = s_{\kappa\nu}^{(\mu)}.$
(ii) The intersection matrices span an algebra $\hat{\mathfrak{C}}[G, X]$ which is isomorphic to the commuting algebra $\mathfrak{C}[G, X]$.

Proof. See [41]. $\square$

Association Schemes.

Definition 6. Let $\mathfrak{A} = \{A_0, A_1, \ldots, A_M\}$ be an incidence scheme. $\mathfrak{A}$ is an *association scheme* if we have in addition

(v) For all $0 \leq i, j \leq M$ holds $A_i A_j = A_j A_i$.

If (iv) is replaced by

(iv') $A_i^+ = A_i$ we call $\mathfrak{A}$ a *symmetric association scheme*.

Commutativity (v) follows already from (i), (ii), (iii), and (iv'). Note that Godsil [44] uses the term 'association scheme' for symmetric association schemes. We use here Delsarte's [45] original terminology.

The theory of association schemes plays a fundamental role in algebraic combinatorics. Many questions concerning distance regular graphs are best discussed in this framework. It plays a fundamental role in coding theory, and is important for the theory of polynomial spaces. For a recent textbook on this topic see [44].

Theorem 7. *The incidence matrices form an association scheme, i.e., the commuting algebra of (G, X) is commutative, if and only if the irreducible constituents of the permutation representation are inequivalent.*

Proof. The argument in [41, (4.10)] is based on [40, Thm. 29.3]. $\qquad\square$

We will not use the incidence matrices $R^{(\mu)}$ themselves but we will rather work with the *symmetrized incidence matrices* $\tilde{R}^{(\mu)}$ defined by

$$\tilde{R}^{(\mu)} = \begin{cases} R^{(\mu)} & \text{if} \quad \mu = \mu^+ \\ R^{(\mu)} + R^{(\mu^+)} & \text{if} \quad \mu \neq \mu^+ \end{cases}$$

Of course, these matrices are again 0-1 matrices. It will be an important question under which conditions they form again an incidence scheme. If this is the case, then they form even a symmetric association scheme which will be denoted by $\tilde{\mathfrak{R}}$. Obviously, a trivial sufficient condition is that $R^{(\mu)} = R^{(\mu^+)}$ for all symmetry classes μ, i.e., if all orbits of G are "self-paired". A less trivial condition can be obtained from the above theorem:

Lemma 8. *If the incidence matrices $R^{(\mu)}$ form an association scheme then the symmetrized incidence matrices form a symmetric association scheme.*

Proof. If the incidence matrices $R^{(\mu)}$ commute, so do the symmetrized incidence matrices $\tilde{R}^{(\mu)}$. It remains to show that $\tilde{R}^{(\mu)}\tilde{R}^{(\nu)}$ is a linear combination of symmetrized incidence matrices. The product of two symmetric matrices is symmetric if and only if they commute, thus the product $\tilde{R}^{(\mu)}\tilde{R}^{(\nu)}$ is again a symmetric matrix. We have

$$\tilde{R}^{(\mu)}\tilde{R}^{(\nu)} = \sum_\kappa p(\kappa)R^{(\kappa)} = (\tilde{R}^{(\mu)}\tilde{R}^{(\nu)})^+ = \sum_\kappa p(\kappa)(R^{(\kappa^+)}).$$

The symmetry of the product implies $p(\kappa) = p(\kappa^+)$, i.e., the product is in fact a linear combination of the symmetrized incidence matrices. $\qquad\square$

2.3 Graphs

Graphs and Their Associated Matrices.

Adjacency Matrix. A graph consists of a set V of N vertices and a set of edges. Two vertices x and y are neighbors (one says that x and y are adjacent) if they are joined by an edge. The set of edges will be denoted by E. We will denote the set of neigbors of a vertex x by $N(x)$. The number of neighbors, $|N(x)|$ is called the *degree* of the vertex x. We will need the diagonal *vertex degree matrix* D which has the entries $D_{xy} \stackrel{\text{def}}{=} |N(x)|\delta_{xy}$, where δ_{ij} is Kronecker's symbol. A graph is *regular* if all vertices have the same degree. In this case we will use D to denote the common vertex degree, i.e., the vertex degree matrix is DI, where I is as usual the identity matrix.

A graph Γ is uniquely described by the $N \times N$ matrix A defined by

$$A_{xy} \stackrel{\text{def}}{=} \begin{cases} 1 & \text{if } y \in N(x) \\ 0 & \text{otherwise.} \end{cases}$$

It is called the *adjacency matrix* of Γ as its non-zero entries correspond to the pairs of adjacent vertices; therefore A is symmetric.

A *path* of length ℓ is a sequence of vertices $x_0, x_1, \ldots, x_\ell$ such that $\{x_{i-1}, x_i\}$ is an edge for $1 \leq i \leq \ell$. A well known result, see, e.g., [46] links the number of paths of given length with the adjacency matrix:

Lemma 9. *The entry x, y in the s-th power of the adjacency matrix, A^s, equals the number of paths of length s from x to y.*

For later reference we note here the following

Definition 10. A *simple random walk* [47] on Γ starting at 0 is a Markov process with transition matrix

$$T = D^{-1}A$$

and initial condition $p_0(x) = \delta_{x,0}$.

Distance Matrices. The notion of a path suggest a natural definition of a distance between two vertices of a graph: let $d(x, y)$ denote the minimum length of a path joining the vertices x and y. In this contribution we will consider only finite connected graphs. Thus $d(x, y)$ is finite for any two vertices. It is easy to check that this distance measure is metric on the vertex set of Γ, i.e., for all $x, y, z \in V$ the following is true:

(i) $d(x, y) = 0$ if and only if $x = y$,
(ii) $d(x, y) = d(y, x)$, and
(iii) $d(x, y) \leq d(x, z) + d(z, y)$.

It is natural to consider the sets

$$N_k(x) \stackrel{\text{def}}{=} \{y \in V \mid d(x, y) = k\}$$

and the *distance matrices* $A^{(k)}$ with entries

$$A_{xy}^{(k)} \stackrel{\text{def}}{=} \begin{cases} 1 & \text{if } d(x, y) = k \\ 0 & \text{otherwise} \end{cases}$$

Of course, $A^{(0)} = I$, the identity matrix, and $A^{(1)} = A$, the adjacency matrix.

Incidence Matrix and Graph Laplacian. A second important matrix links the vertices and edges of Γ. For each edge $h = \{v, w\}$ we choose one of the two vertices as the "positive end" and the other one as the "negative end" of the edge. The choice of the orientation is completely arbitrary. The matrix

$$\nabla_{ij}^+ = \begin{cases} +1 & \text{vertex } v_i \text{ is the positive end of edge } e_j \\ -1 & \text{vertex } v_i \text{ is the negative end of edge } e_j \\ 0 & \text{otherwise} \end{cases}$$

is called the *incidence matrix of* Γ. The notion of an incidence matrix of a graph should not be confused with the incidence matrices of transitive permutation groups described above — they are unrelated. The choice of the symbol ∇ is intentional. In fact, let $f : V \to \mathbb{R}$ be an arbitrary function. Then $(\nabla f)(h) = f(v) - f(w)$ where h is the edge $\{v, w\}$, and v is the positive end of the edge h. This is as close to a differential operator as one can get on a graph.

The matrix

$$\Delta = A - D$$

is called the *Laplacian* on Γ. For regular graphs this becomes $A - DI$, where D is now the constant vertex degree of Γ. For an arbitrary function $f : V \to \mathbb{R}$ we have

$$(\Delta f)(x) = \sum_{y \in N(x)} \big(f(y) - f(x) \big).$$

The graph Laplacian shares its most important properties with the familiar Laplacian operator in $\mathbb{R}^n$ as listed in the following

Theorem 11. *The graph Laplacian Δ has the following properties:*

(i) Δ *is symmetric.*
(ii) Δ *is non-positive definite.*
(iii) Δ *is singular; the eigenvector* $(1, \ldots, 1)$ *belongs to the eigenvalue* $\tilde{\Lambda}_0 = 0$*. If Γ is connected (as we will always assume), then Λ_0 has multiplicity 1.*
(iv) $\Delta = -\nabla^+ \nabla$*, that is, it corresponds to "second derivatives" on the graphs.*
(v) *for all landscapes f and g holds Green's formula in the following form*

$$\sum_{x \in V} f(x)(\Delta g)(x) = \sum_{x \in V} g(x)(\Delta f)(x) = - \sum_{h \in E} (\nabla f)(h)(\nabla g)(h).$$

(vi) $\langle \nabla f, \nabla g \rangle = -\langle f, \Delta g \rangle = -\langle g, \Delta f \rangle$ *for arbitrary* $f, g : V \to \mathbb{R}$*, where* $\langle . , . \rangle$ *is the usual scalar product in* $\mathbb{R}^N$.

Proof. (i) is obvious, (ii) and (iii) are well known, see, e.g., [46, 48], and (iv) is Proposition 4.8 of [46]. Green's formula (v) is checked by explicit calculation, see e.g., [49].   □

The graph Laplacian has received considerable attention in the theory of electrical networks, see, e.g., [46, Chap.5]. A recent book on potential theory on infinite discrete lattices is [50]. Mark Kac [51] has asked whether the knowledge of the spectrum of a Laplacian with Dirichlet boundary conditions is sufficient to completely determine the shape of its domain. Shortly after Kac's talk a series of papers, e.g., [52, 53, 54] investigated the analogous question for the graph Laplacian: Given the spectrum of Δ, how much geometric information on the underlying graph can be retrieved?

The set of eigenvalues of the adjacency matrix is called the spectrum of the graph Γ. The spectral properties of the matrices A, T, and Δ are closely related. For later reference we note the following trivial

Lemma 12. *Let* Γ *be a regular graph with adjacency matrix* A *and vertex degree* D*. Then* A*,* T*, and* Δ *have the same eigenvectors and the corresponding eigenvalues* Λ_k*,* λ_k*, and* $\bar{\Lambda}_k$ *belonging to the eigenvector* ϕ_k *are related via*

$$\bar{\Lambda}_k = \Lambda_k - D \qquad and \qquad \lambda_k = \Lambda_k / D.$$

Symmetry and Regularity of Graphs. An *automorphism* of Γ is mapping $\alpha : V \to V$, such that $(\alpha(x), \alpha(y))$ is an edge of Γ if and only if (x, y) is an edge of Γ. The set of all automorphisms forms a group, the automorphism group **Aut**$[\Gamma]$.

A large number of different symmetry properties have been defined for graphs. These notions can be subdivided into two classes depending on whether their definition uses properties of the automorphism group. For reference we mention here the most important ones

Regularity Conditions.

- Γ is *regular* if all vertices have the same number D of neighbors.
- Γ is *distance degree regular* if the cardinality of the sets

$$N_k(x) \stackrel{\text{def}}{=} \{ y \in V \mid d(x, y) = k \}$$

  depends only on k.
- Γ is *walk regular* if the number of closed walks of length r starting at vertex x is independent of x for all r, or equivalently, if $(A^r)_{xx}$ is independent of x for all r.

- Γ is *distance regular* if the cardinalities of the intersections $N_i(u) \cap N_j(v)$ depends only on the distance $d(u,v)$. The distance matrices $A^{(k)}$ with entries $a_{xy}^{(k)} = 1$ if $d(i,j) = k$ and $a_{xy}^{(k)} = 0$ otherwise form a symmetric association scheme.
- Γ is *strongly regular* if the cardinality of the intersections $N_1(u) \cap N_1(v)$ depends only on whether u and v are adjacent or not. Usually one excludes complete graphs. Strongly regular graphs are then distance regular with diameter 2. They play a prominent role in algebraic combinatorics.
- Γ is *k-set-regular* if, given a subset Y consisting of at most k vertices, the number of vertices joined to each vertex in Y depends only on the isomorphism type of the induced subgraph of Y in Γ. Consequently, 1-set-regular is regular, and 2-set-regular is strongly regular.

Transitivity Conditions.

- Γ is *vertex transitive* if $\mathbf{Aut}[\Gamma]$ acts transitively on V, i.e., if for any two vertices x and y there is an automorphism α such that $\alpha(x) = y$.
- Γ is *generously vertex transitive* if $\mathbf{Aut}[\Gamma]$ acts generously transitively on V, i.e., if for any two vertices x and y there is an automorphism α such that $\alpha(x) = y$ and $\alpha(y) = x$.
- Γ is *edge transitive* if $\mathbf{Aut}[\Gamma]$ acts transitively on the set E of edges (considered as unordered pairs of vertices).
- Γ is *weakly symmetric* if it is both vertex transitive and edge transitive.
- Γ is *symmetric* if for all vertices $x, y, u, v \in V$ such that (x,y) and (u,v) are edges there is an automorphism α fulfilling $\alpha(x) = u$ and $\alpha(y) = v$.
- Γ is *t-arc transitive* if $\mathbf{Aut}[\Gamma]$ acts transitively on "arcs" consisting of at most t subsequent edges such that two consecutive edges in the arc are not identical. Note that arcs can contain a vertex or an edge twice, but it is not allowed to backtrack a single step. 1-arc-transitive is the same as "symmetric".
- Γ is *distance transitive* if for any four vertices $x, y, x', y' \in V$ such that $d(x,y) = d(x',y')$ there is an automorphism fulfilling $\alpha(x) = x'$ and $\alpha(y) = y'$.
- Γ is *metrically k-transitive* if for any two k-tuples of vertices $(x_1, \ldots, x_k)$ and $(y_1, \ldots, y_k)$ which satisfy $d(x_i, x_j) = d(y_i, y_j)$ there is an automorphism fulfilling $\alpha(x_i) = y_i$. In other words, a graph is metrically k-transitive if any isometry between sets of at most k vertices can be extended to an automorphism. Metrically 1-transitive is the same as vertex-transitive, metrically 2-transitive is the same as distance-transitive.
- Γ is *k-set-transitive* if any isomorphism between induced subgraphs of sets consisting of at most k vertices extends to an automorphism of Γ. For $k \geq 2$, k-set transitive graphs are metrically transitive with diameter 2.
- Γ is *homogeneous* if it is k-set transitive for all k.
- Γ is *metrically homogeneous* if it is metrically k-transitive for all k.

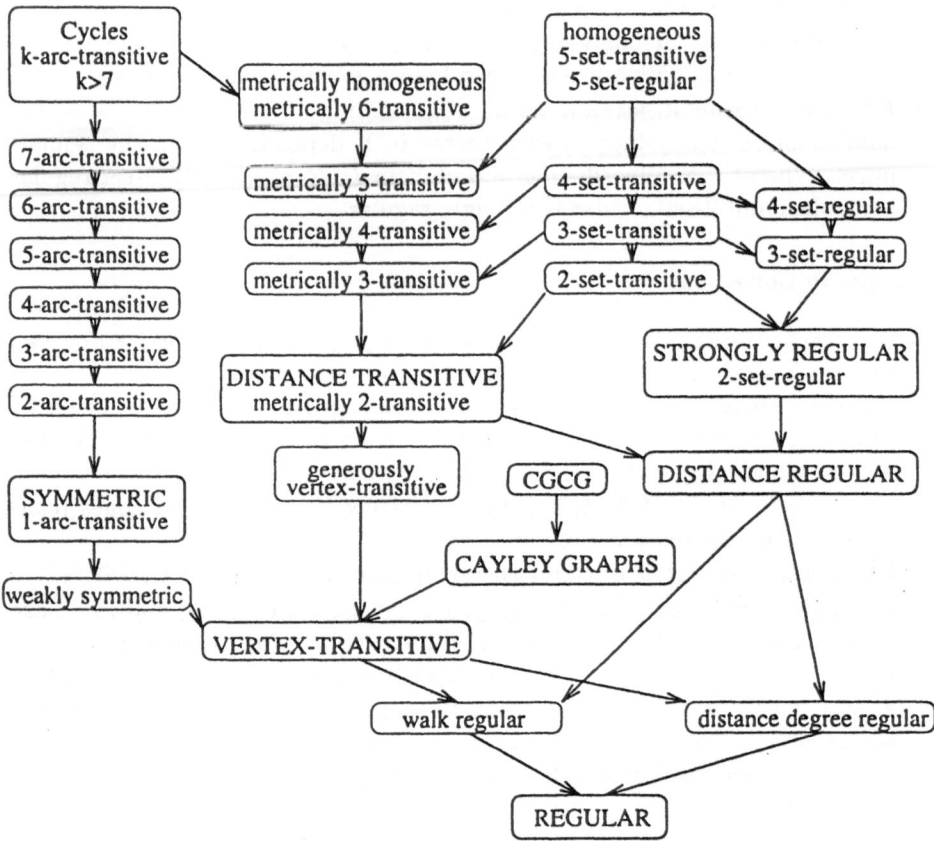

Fig. 2. A zoo of symmetry properties of graphs has been studied. Arrows indicate implications; equivalent properties are shown within the same rectangle. CGCG refers to Cayley graphs with an Abelian group. Cayley graphs are discussed in section 2.3.4.

The relations between the animals in this zoo of graph theoretical properties is shown in figure 2. In the following sections we will discuss the most important properties in a little more detail.

Vertex-Transitive Graphs. A most important class of graphs are derived from groups: Let H be a group with N elements and Φ a set of generators[3] of H such that

(i) $e \notin \Phi$, and
(ii) $x \in \Phi$ implies x^{-1} in Φ.

Then the graph $\Gamma(H, H', \Phi)$ with vertex set $V = H$ and edge set $\{(x, y) | xy^{-1} \in \Phi\}$ is called the CAYLEY *graph* of the group H with respect to the set of generators Φ.

Not all vertex transitive graphs are Cayley graphs, however. A famous counterexample is the Petersen graph, figure2.

Sabidussi [55] gives a method for constructing all vertex transitive graphs from groups which is closely related to Cayley graphs. Let $\sim$ be an equivalence relation on the vertex set of a graph Γ. Then the graph $\Gamma/_\sim$ has the equivalence classes as its vertices, and two such vertices are joined by an edge if and only if there is at least one edge connecting the two classes in Γ. Now let $H' < H$ be a subgroup of H, and let Φ again be a set of generators of H. The left cosets of H' induce an equivalence relation $\sim$ on H. We will denote the graph $\Gamma(H, \Phi)/_\sim$ by $\Gamma(H, H', \Phi)$. Explicitly, its vertex set and its edge set are

$$V = \{gH' | g \in H\} \quad \text{and} \quad E = \{(aH', bH') | aH' \neq bH' \wedge aH' \cap (bH'\Phi) \neq 0\}.$$

As an example consider the Cayley graph of the symmetric group S_n with the set of all transpositions as set of generators. The set of all rotations forms a cyclic group $\mathbb{Z}_n < S_n$. The graph $\Gamma(S_n, \mathbb{Z}_n, T)$ is a sensible choice for dealing with a traveling salesman problem, see sect. 2.1.

Theorem 1 *A graph Γ is vertex transitive if and only if there is a group H, a subgroup $H' < H$ and a set of generators Φ fulfilling (i) and (ii) above with $\Phi \cap H' = \emptyset$ such that $\Gamma = \Gamma(H, H', \Phi)$.*
(2) A graph Γ is a Cayley graph if and only if its automorphism group contains a subgroup G which acts regularly on the vertex set. Then $\Gamma = \Gamma(G, \Phi)$ with an appropriate set of generators.

Proof. The first statement is due to Sabidussi [55], the second one is proved in [56, 46]. $\square$

Later on we will need a few properties of simple random walks on vertex transitive graphs. If Γ is vertex transitive, let us define the probability $\varphi_{s\mu}$ that a simple random walk of s steps ends in a vertex contained in the symmetry class μ.

An explicit expression for $\varphi_{s\mu}$ is given in [57], see sect. 4.1.2. The relation of $\varphi_{s\mu}$ and the powers of the adjacency matrix is also of interest: for vertex

---

[3] Φ is a set of generators of the group H if each element $x \in G$ can be represented as a product of members of H.

transitive graphs one finds [49]

$$A^s = \sum_\mu \frac{D^s}{|\mu|} \varphi_{s\mu} R^{(\mu)} \stackrel{\text{def}}{=} \sum_\mu \Theta_{s\mu} R^{(\mu)}.$$

Cayley Graphs. It is easy to see that a left multiplication of $x \in G$ by an arbitrary $g \in G$, i.e., the permutation action $x \mapsto gx$, is an automorphism of $\Gamma(G, \Phi)$, i.e., the group of left multiplications $\tilde{G}$ is a subgroup of $\mathbf{Aut}[\Gamma(G, \Phi)]$. By construction $\tilde{G}$ acts transitively on G. Consequently, (yz, y) belongs to the same orbit of $\tilde{G}$, and hence also to the same orbit of $\mathbf{Aut}[\Gamma(G, \Phi)]$, as (z, e), where e is the group identity which we choose as reference vertex in the Cayley graph. Setting $x = yz$ this implies

$$xy^{-1} = z \Longrightarrow (x, y) \in \ddot{\mu}_z$$

where μ_z denotes the symmetry class to which z belongs. Now consider the $N \times N$ matrices defined by

$$G_{xy}^{(z)} = \begin{cases} 1 & \text{if } xy^{-1} = z \\ 0 & \text{otherwise} \end{cases}$$

Lemma 13. *The matrices $G^{(z)}$ form an incidence scheme $\mathfrak{G}$ with the following properties:*

(i) $G^{(e)} = I$.
(ii) $\sum_{z \in G} G^{(z)} = J$.
(iii) $G^{(r)} G^{(s)} = G^{(rs)}$.
(iv) $G^{(z^{-1})} = (G^{(z)})^+$.

Proof. (i) and (ii) are obvious. Properties (iii) and (iv) are checked by direct computation.   □

The algebra spanned by $\mathfrak{G}$ is (isomorphic to) the *group algebra* of G. In particular, if G is a commutative group, then $\mathfrak{G}$ is an association scheme, and $\langle \mathfrak{G} \rangle$ is a commutative algebra.

The matrices $R^{(\mu)}$ are obtained from the matrix representation of the group algebra via

$$R^{(\mu)} = \sum_{z \in \mu} G^{(z)} \qquad \text{and} \qquad \tilde{R}^{(\mu)} = \sum_{z \in (\mu \cup \dot{\mu}^+)} G^{(z)}.$$

An immediate consequence of this representation is the following

Lemma 14. *Let Γ be a Cayley graph of a commutative group G. Then the symmetrized incidence matrices $\tilde{R}^{(\mu)}$ form a symmetric association scheme.*

Proof. It suffices to observe that the matrices $R^{(\mu)}$ commute. The lemma then follows immediately from the lemma in sect. 2.2.5.   □

The (Cartesian) product $\Gamma_1 \times \Gamma_2$ of two graphs has vertex set $V(\Gamma_1 \times \Gamma_2) = V(\Gamma_1) \times V(\Gamma_2)$. Two vertices (x_1, x_2) and (y_1, y_2) are connected by an edge if either (i) $x_1 = y_1$ and x_2, y_2 are adjacent in Γ_2, or (ii) $x_2 = y_2$ and x_1, y_1 are adjacent in Γ_1. It is easily checked that the product of two Cayley graphs $(G_1, \Phi_1) \times (G_2, \Phi_2)$ is the Cayley graph $(G_1 \times G_2, \Phi)$, where $G_1 \times G_2$ is the direct product of the two groups and $\Phi = (\Phi_1 \times \{e_2\}) \cup (\Phi_2 \times \{e_1\})$, e_1 and e_2 being the group identities.

Distance Regular Graphs. A connected graph Γ is *distance regular* if for any two vertices u and v, and any two integers k and l, the numbers

$$s_{kl}^{(d)} \overset{\text{def}}{=} |N_k(u) \cap N_l(v)|$$

depend only on $d \overset{\text{def}}{=} d(u, v)$, that is, on the distance of the vertices u and v. The numbers $s_{kl}^{(d)}$ are known as the intersection numbers of Γ. Let x by an arbitrary reference vertex. Then

$$\varpi = (\{x\}, N(x), N_2(x), \dots, N_M(x))$$

is called the *distance partition* of Γ centered at x, where $M = \text{diam}\Gamma$, the diameter of the graph. Distance regularity strongly restricts both the algebraic and the geometric properties of a graph. The most important features are summarized below.

Lemma 15. *Let Γ be distance regular. Then the distance matrices $A^{(d)}$ form a symmetric association scheme, and $A = A^{(1)}$ has exactly $M + 1$ distinct eigenvalues.*

Proof. See, e.g., [44, Chap.11]. □

Lemma 16. *The sequence $k_r \overset{\text{def}}{=} |N_r(x)|$ is unimodal and $k_r \leq k_{M-r}$ for all $r \leq M/2$. Thus the average distance $\bar{d}$ of two vertices is at least $\text{diam}\Gamma/2$.*

Proof. See, e.g., [44, Chap.11]. □

Distance transitivity implies distance regularity. The converse is not true. A counterexample can be found in [58], see also [46, 20c].

2.4 Fourier Series on Graphs

A series expansion in terms of a complete and orthonormal system of eigenfunctions of the Laplace operator is commonly termed *Fourier expansion*. We will adopt the same terminology here following Weinberger's paper [59]. Thus let $f : V \to \mathbb{R}$ be a real valued function on the vertex set of Γ and let $\{\theta_s\}$ denote a complete orthonormal set of eigenvectors of the graph Laplacian Δ. Then we call

$$f(x) = \sum_{y \in \Gamma} b_y \theta_y(x)$$

the *Fourier expansion* of f.

It will turn out to be convenient to label the eigenvectors θ_y by the vertices of the underlying graph Γ. This is possible because the eigensystem of the finite symmetric operator Δ is complete. In general, this labeling is arbitrary.

For Cayley graphs of commutative groups (CGCGs), however, the eigenvectors and eigenvalues have a particularly simple form. Let $x = (x_1, \ldots, x_m)$ denote the component wise representation of the group element x which is obtained from the decompositions of the commutative group G into a direct product of m cyclic groups of orders N_k. The group action is represented by component-wise addition modulo N_k:

$$x \circ y = (x_1 + y_1 \bmod N_1, x_2 + y_2 \bmod N_2, \ldots, x_m + y_m \bmod N_m).$$

The Fourier basis of a CGCG, which coincides with a complete set of eigenvectors since Cayley graphs are regular, are of the form [60]

$$e_g(x) = \exp\left(2\pi i \sum_k \frac{x_k g_k}{N_k}\right),$$

and the corresponding eigenvalues are given by

$$\Lambda_g = \sum_{x \in \Phi} e_g(x) \qquad \text{and} \qquad \bar{\Lambda}_g = \sum_{x \in \Phi} (e_g(x) - 1).$$

Note that in this special case we have a canonical labelling of the eigenvectors. For convenience we recall a few basic properties of the functions $e_x(y)$.

(i) $e_g(x) = e_x(g)$.
(ii) $e_g(x \circ y) = e_g(x)e_g(y)$ and $e_{g \circ h}(x) = e_g(x)e_h(x)$.
(iii) $e_g(x)e_g^*(x) = 1$, where the asterisk denotes complex conjugation.
(iv) $\sum_{x \in G} e_g(x)e_h^*(x) = N\delta_{gh}$ and $\sum_{l \in G} e_l(x)e_l^*(y) = N\delta_{xy}$, where N is the total number of configurations.

These functions have the well-known form of the Fourier basis on $\mathbb{C}^n$. They are, of course, the characters of the group commutative G. Except for property (ii), which will be crucial for some results in the following, the above statements are immediate consequences of Schur's Lemma, and are therefore true for the characters of arbitrary groups, see, e.g., [61].

Since Δ is symmetric, the Fourier basis $\{\theta_y\}$ can always be chosen real valued; it is convenient, however, to allow for complex conjugate pairs of eigenvectors. In particular, we will use the following identity for functions on a CGCG:

$$f(x) = \sum_{l \in G} a_l e_l(x) = \sum_{l \in G} a_l^* e_l^*(x).$$

Since we deal with a finite vector space with a scalar product (for which we will use the notation $\langle ., . \rangle$ in this paper), spanned by eigenvectors $\{\theta_y\}$ of the graph Laplacian, the familiar properties of Fourier series, such as *Parseval's equation*

$$\|f\|^2 = \langle f, f \rangle = \sum_{y \in \Gamma} \left(\frac{\langle f, \theta_y \rangle}{\langle \theta_y, \theta_y \rangle} \right)^2 ,$$

and the *mean square approximation theorem* hold for all landscapes on all connected graphs. For the convenience of the reader we recall

Proposition 17 *Mean Square Approximation Theorem. Consider a landscape f on Γ with Fourier expansion*

$$f(x) = \sum_{y \in V} a_y \theta_y(x)$$

Let X be a subset of V, let

$$g(x) = \sum_{y \in X} b_y \theta_y(x)$$

be an approximation to f. Then the squared approximation error $\|f - g\|^2 = \langle (f - g), (f - g) \rangle$ is minimized by choosing $b_y = a_y \equiv \langle f, \theta_y \rangle$ for all $y \in X$.

2.5 Spectral Properties of the Adjacency Matrix

Equitable Partitions.

Definition 18. Let $\varpi = (\varpi_0, \varpi_1, \ldots, \varpi_M)$ be a partition of the vertex set V into the *cells* ϖ_i, $i = 0, \ldots, M$. ϖ is called *equitable* if the number of neighbors which a vertex in ν has in cell μ is independent of the choice of the vertex in ν. In other words, ϖ is equitable if for all $\mu, \nu \in \varpi$ holds

$$\breve{A}_{\mu\nu} \stackrel{\text{def}}{=} |N(y) \cap \mu| = \sum_{x \in \mu} A_{xy} . \quad \text{for all } y \in \nu.$$

We call $\breve{A}$ a *combinatorially collapsed adjacency matrix*.

There is an $(M + 1) \times N$ matrix associated with each partition ϖ of V into $M + 1$ cells. This matrix will also be denote by ϖ:

$$\varpi_{\mu x} \stackrel{\text{def}}{=} \begin{cases} 1 & \text{if } x \in \mu \\ 0 & \text{otherwise} \end{cases}$$

We remark that this definition is the transpose of the convention in Godsil's book [44], while it conforms the notation used by Bollobás [43]. Equitable partitions have been introduced by Schwenk [62]; more recently they have been used by Powers and coworkers as *colorations*, see, e.g., [63, 64], see also [65, Chap.4].

Theorem 19. *Let ϖ an equitable partition of the vertex set V of a graph Γ. Then the following statements hold.*

(i) $\varpi A = \breve{A}\varpi$.

(ii) If e is an eigenvector of A with eigenvalue Λ, then $v = \varpi e$ is a right eigenvector of $\breve{A}$ with the same eigenvalue provided $v \neq 0$. For later reference we note the explicit formula

$$v(\mu) = \sum_{x \in \mu} e(x)$$

(iii) If v is a right eigenvector of $\breve{A}$, then

$$\rho(\mu) = \frac{1}{|\mu|} v(\mu)$$

is a left eigenvector of $\breve{A}$.

(iv) If ρ is a left eigenvector of $\breve{A}$, then $\rho\varpi$ is an eigenvector of A which is constant on all cells.

(v) $|\mu|(\breve{A}^r)_{\mu\nu} = |\nu|(\breve{A}^r)_{\nu\mu}$.

(vi) The characteristical polynomial of $\breve{A}$ divides the characteristical polynomial of A.

(vii) If Γ is connected then the A and $\breve{A}$ have the same spectral radius.

Proof. See [44, chap.5]. $\square$

There is an intimate relation between the orbits of a group of automorphisms acting on V and the equitable partitions of a graph.

Lemma 20. Let $G \leq Aut[\Gamma]$ be an arbitrary group of automorphisms of Γ. Then the orbits of G form an equitable partition ϖ_G of V.

Proof. See [44, p.76]. $\square$

The converse is not true. There are equitable partitions which are not induced by any automorphism group. For an example see [44, p.76]. One says that a partition ϖ' is a *refinement* of ϖ, in symbols $\varpi' \leq \varpi$, if each cell of ϖ' is contained in some cell of ϖ. The equitable partitions form a lattice with respect to refinement [44, p.87].

Equitable Partitions with a Reference Vertex. The results (vi) and (vii) on equitable partitions can be strenghtened if ϖ fulfills a single, simple additional condition. An equitable partition ϖ which has a cell $\varpi_0 = \{u\}$ consisting of the single vertex u will be called *equitable partition with reference vertex*.

Theorem 21. Let ϖ be an equitable partition with reference vertex u. Then A and $\breve{A}$ have the same minimal polynomial, and Λ is an eigenvalue of A if and only if it is an eigenvalue of $\breve{A}$.

Proof. See [43, Thm.8.6]. $\square$

As an example we note that the distance partitions of a distance regular graph as well as the orbits of an arbitrary stabilizer subgroup $G_x < \text{Aut}[\Gamma]$ of a vertex transitive graph form an equitable partition of V with a reference vertex.

Lemma 22. *Let ϖ be an equitable partition of a graph Γ with reference vertex 0. Then there is always a basis $\{\rho_i\}$ of left eigenvectors of the corresponding combinatorially collapsed adjacency matrix $\check{A}$ with the following properties:*

(i) The eigenvectors are normalized such that $\rho(0) = v(0) = 1$.
(ii) The left eigenvectors are orthogonal w.r.t. the scalar product
$$\langle a; b \rangle = \sum_\mu a(\mu) b(\mu) |\mu|.$$
(iii) This basis is unique if all eigenvalues of $\check{A}$ are simple.

Proof. The corresponding result in [57] is formulated for the collapsed adjacency matrix $\hat{A}$ of a vertex transitive graph, but only the properties of equitable partitions with a reference vertex are used in the course of the proof. □

A few remarks on distance regular graphs are in order here.

Lemma 23. *Let Γ be distance regular and let ϖ be a distance partition with respect to an arbitrary reference vertex x. Then $\check{A}_{kl}^{(d)} = s_{kl}^{(d)}$.*

Proof. This follows immediately from the discussion in [44, Chap.11]. □

The lemma states that we recover the intersection numbers of a distance regular graph as the combinatorially collapsed distance matrices. This is analogous to the relation of the incidence matrices and their intersection numbers in Higman's theory, although there are no group actions involved in the case of distance regular graphs.

The Adjacency Algebra. The *adjacency algebra* of a graph Γ is the matrix algebra spanned by the powers of the adjacency matrix. It will be denoted by $\mathfrak{A}[\Gamma]$. Of course, this algebra is commutative. It is clear from the previous section that the properties of $\mathfrak{A}[\Gamma]$ are closely linked to the equitable partitions with reference vertices, since its dimension coincides with the degree of the minimal polynomial of A or $\check{A}$. In particular, therefore, the dimension of $\mathfrak{A}[\Gamma]$ is bounded above by 1 plus the number of cells in an equitable partition with reference vertex. On the other hand we have the following lower bound on the dimension of the adjacency algebra:

Theorem 24. *Let Γ be a connected graph with diameter $\text{diam}\,\Gamma$. Then the dimension of the adjacency algebra $\mathfrak{A}[\Gamma]$ is at least $\text{diam}\,\Gamma + 1$.*

Proof. See [46, Prop.2.6]. □

This result suggests that equitable partitions with a reference vertex x are always refinements of the corresponding distance partition centered at the reference vertex x. We do not know whether this is actually true. Another important question is whether there is always an equitable partition with reference vertex such that all eigenvalues of $\breve{A}$ are simple. The above discussion, along with the striking similarity of the algebraic properties of distance regular graphs and the theory of incidence structures seems to suggest that it might even be possible to base the investigation of configuration spaces entirely on equitable partitions with reference vertices instead of graph automorphisms.

Here we will return, however, to the safer grounds of vertex transitive graphs, for which Higman's beautiful theory of intersection numbers is applicable, since the automorphism group of a vertex transitive graph acts by definition transitively on V. The incidence scheme $\mathfrak{R}[\Gamma]$ associated with a vertex transitive graph Γ is of course the one of $\mathbf{Aut}[\Gamma]$ acting on $V \times V$, see sect. 2.2. The algebra spanned by $\mathfrak{R}[\Gamma]$ is usually referred to as the *commuting algebra of the graph*, denoted by $\mathfrak{C}[\Gamma]$. Since the adjacency matrix A is a linear combination of the matrices $R^{(\mu)}$ we have necessarily $\mathfrak{A}[\Gamma] \subseteq \tilde{\mathfrak{R}}[\Gamma]$.

Theorem 25. *Let Γ be vertex transitive. Then the following statements are equivalent:*

(i) $\mathfrak{A}[\Gamma] = \mathfrak{C}[\Gamma]$.
(ii) All eigenvalues of the collapsed adjacency $\hat{A}$ matrix are simple.
(iii) The powers of A span $\mathfrak{C}[\Gamma]$, i.e., all matrices $R^{(\mu)}$ are polynomials in A.
(iv) The powers of $\hat{A}$ span $\hat{\mathfrak{C}}[\Gamma]$.

Proof. (i) $\Longleftrightarrow$ (ii) $\Longleftrightarrow$ (iv) follows immediately from [41, (4.12)].
(i) $\Longleftrightarrow$ (iii)   The matrices $R^{(\mu)}$ form the basis of a vector space of dimension $M + 1$. We have seen in section 2.3.3 that

$$A^s = \sum_{\mu} \Theta_{s\mu} R^{(\mu)} \qquad \text{with} \qquad \Theta_{s\mu} = \frac{D^s \varphi_{s\mu}}{|\mu|}.$$

The coefficients $\Theta_{s\mu}$ form a $(M + 1) \times (M + 1)$ matrix which can be invertible only if the number of distinct eigenvectors of A equals the number of symmetry classes, i.e., if all eigenvalues of the collapsed adjacency matrix are simple. Since $\mathfrak{A}[\Gamma] \subseteq \mathfrak{C}[\Gamma]$ the converse is also true. Hence Θ is invertible if and only if all eigenvalues of $\hat{A}$ are simple. $\square$

Property (iii) has been termend " Θ-property" in [49]. It seems to be of particular importance for the theory of landscapes and random fields. It is not known (to the author) whether the Θ-property is related to any of the well-studied graph theoretical properties of configuration spaces. Not all vertex transitive graphs fulfill (iii); An example for a vertex transitive graph without the Θ-property is the permutohedron, i.e., the Cayley $\Gamma(S_n, \mathcal{K})$ of the symmetric group with the set of canonical transpositions $(i, i + 1)$ as generators.

2.6 Examples of Configuration Spaces

Sequence Spaces. Consider the set of all sequences of length n, which are composed of letters taken from some finite alphabet of size α. The canonical distance is given by the number of positions in which two sequences x and y differ; this metric distance measure is the Hamming distance [66]. The graph obtained from connecting two sequences if and only if their Hamming distance is $d(x, y) = 1$ is called the *Hamming graph* or *sequence space* $\mathcal{Q}_\alpha^n$. The sequence space $\mathcal{Q}_\alpha^n$ is the n-fold product of the complete graph with α vertices with itself. A complete graph with α vertices can be regarded as the Cayley graph graph of an arbitrary group G of order α with $\Phi = G \setminus \{e\}$. Thus Hamming graphs are Cayley graphs of commutative groups, and they are distance transitive, see e.g., [46].

Hamming graphs constitute the most common class of configuration spaces. Their automorphism group is well known [67]

$$\mathbf{Aut}[\mathcal{Q}_\alpha^n] = S_\alpha \wr S_n = S_n \vert \times (S_\alpha)^n \,,$$

where S_m denotes the symmetric group on m elements.

The following results on the eigenvalues and eigenvectors of the sequence spaces are well known, see, e.g., [68, 69, 70, 71, 72, 73, 74].

The eigenvalues and left eigenvectors of the collapsed adjacency matrix $\hat{A}$ of a sequence space $\mathcal{Q}_\alpha^n$ are given by

$$\Lambda_p = n(\alpha - 1) - p\alpha,$$

$$\rho_p(d) = \frac{1}{(\alpha - 1)^p \binom{n}{p}} \mathbf{K}_p^{(\alpha,n)}(d),$$

where $\mathbf{K}_p^{(\alpha,n)}$ are the *Krawtchouk polynomials*

$$\mathbf{K}_p^{(\alpha,n)}(d) = \sum_{j=0}^{p} (-1)^j \binom{d}{j} \binom{n-d}{p-j} (\alpha - 1)^{p-j}.$$

The multiplicity of Λ_p w.r.t. the full adjacency matrix A is $m(\Lambda_p) = (\alpha - 1)^p \binom{n}{p}$; all eigenvalues are simple w.r.t. the collapsed adjacency matrix $\hat{A}$. The left eigenvectors ρ_p of $\hat{A}$ fulfill the orthogonality relation

$$\langle \rho_k; \rho_l \rangle = \frac{\kappa^n}{(\kappa - 1)^k \binom{n}{k}} \delta_{kl}$$

with respect to the scalar product

$$\langle a; b \rangle = \sum_{d=0}^{n} (\kappa - 1)^d \binom{n}{d} a(d) b(d),$$

and the symmetry relation $\rho_p(d) = \rho_d(p)$.

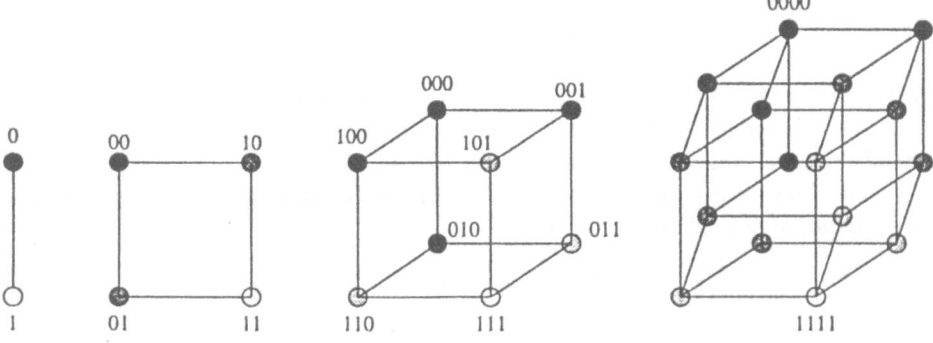

Fig. 3. Hypercubes Q_2^n for $n = 1$ through $n = 4$. Vertices belonging to the same symmetry class μ are marked by a common symbol. Vertices within a symmetry class have common distance from the reference vertex ●. The hypercubes are distance transitive, since their symmetry classes are defined by the distance from the reference vertex.

Generalized Hamming Graphs. The name *Generalized Hamming Graph* is sometimes used for direct products of sequence spaces. They correspond to sequences where the alphabet is not the same for all positions. Such graphs have turned out to be useful when considering the inverse folding problem for RNA secondary structures [75, 76]: An RNA secondary structure can be partioned into the unpaired regions where each position can be realized by one of the 4 nucleotides **G**, **C**, **A**, and **U**, and the paired regions, where each base pair can be realized by one of the 6 types of base pairs (**GC**, **CG**, **AU**, **UA**, **GU**, and **GU**).

The set of sequences compatible with the given structure can then be considered as the graph

$$\mathcal{C} \cong \mathcal{Q}_4^{n_u} \times \mathcal{Q}_6^{n_p},$$

where n_u and n_p are the numbers of the unpaired positions and of the base pairs, respectively.

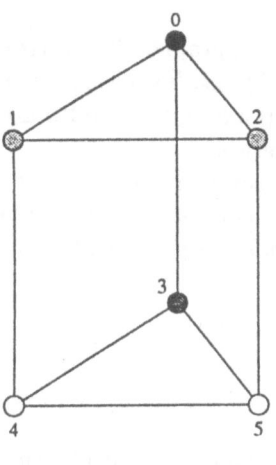
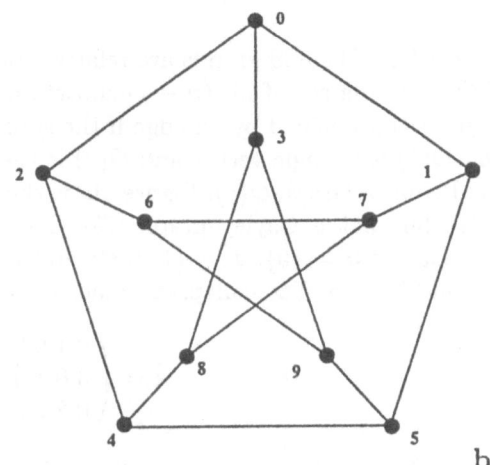

a   b

Fig. 4. a) The trigonal prism graph is the smallest vertex transitive graph which is not edge transitive. Its vertex set decomposes into four symmetry classes which are the reference vertex $\mathbf{0} = \{0\}$ itself, its neighbors within the same triangle $\mathbf{1a} = \{1, 2\}$, its neighbor in the other triangle $\mathbf{1b} = \{3\}$, and the remaining vertices $\mathbf{2} = \{4, 5\}$ which have distance two from the reference vertex.
b) The Petersen graph is the smallest distance transitive graph which is not a Cayley graph of any group. It has three classes of vertices: the reference vertex $\mathbf{0} = \{0\}$, its neighbors $\mathbf{1} = \{1, 2, 3\}$, and all other vertices $\mathbf{2} = \{4, 5, 6, 7, 8, 9\}$.

The simplest case of a Generalized Hamming Graph which is not a sequence space is the trigonal prism $\mathcal{Q}_2 \times \mathcal{Q}_3$, see figure 4a. Its collapsed adjacency matrix is

$$\hat{A} = \begin{pmatrix} 0 & 1 & 1 & 0 \\ 2 & 1 & 0 & 1 \\ 1 & 0 & 0 & 1 \\ 0 & 1 & 2 & 1 \end{pmatrix}$$

The eigenvalues of $\hat{A}$, the corresponding *left* eigenvectors, and the multiplicities of these eigenvalues in A are

$$
\begin{aligned}
&\Lambda_0 = 3 && \rho_0 = (1, 1, 1, 1) && m(\Lambda_0) = 1 \\
&\Lambda_1 = 1 && \rho_1 = (1, 1, -1, -1) && m(\Lambda_1) = 1 \\
&\Lambda_2 = 0 && \rho_2 = (1, -\tfrac{1}{2}, 1, -\tfrac{1}{2}) && m(\Lambda_2) = 2 \\
&\Lambda_3 = -2 && \rho_3 = (1, -\tfrac{1}{2}, -1, \tfrac{1}{2}) && m(\Lambda_3) = 2
\end{aligned}
$$

Johnson Graphs. The vertices of the Johnson graph $\mathcal{J}(n,k)$ are the subsets consisting of exactly k elements from a set with n elements. Two vertices are adjacent if the corresponding subsets have $k-1$ exactly elements in common. The Johnson graphs are distance transitive, see, e.g., [44]. The graph $\mathcal{J}(n,n/2)$ for even n is the natural configuration spaces for bipartitioning problems, see sect. 4.4.2. [77, 78, 44].

Odd Graphs. The odd graphs are relatives of the Johnson graphs. The vertex set of $O(n)$ is the set of all $(n-1)$-subsets of a $(2n-1)$ set. Two vertices x and y in $O(n)$ are joined by an edge if the subsets corresponding to x and y are disjoint. $O(1)$ is a single vertex and $O(2)$ is the triangle graph.

The Petersen Graph $O(3)$, Figure 4b, is the smallest graph which is distance transitive but not a Cayley graph [79]. The symmetry classes are the three distance classes $\mathbf{0} = \{0\}$, $\mathbf{1} = \{1,2,3\}$, and $\mathbf{2} = \{4,5,6,7,8,9\}$, with $|\mathbf{0}| = 1$, $|\mathbf{1}| = 3$, and $|\mathbf{2}| = 6$. The collapsed adjacency matrix is

$$\hat{A} = \begin{pmatrix} 0 & 1 & 0 \\ 3 & 0 & 1 \\ 0 & 2 & 2 \end{pmatrix}$$

The eigenvalues of $\hat{A}$, the corresponding *left* eigenvectors, and the multiplicities of these eigenvalues in A are

$$
\begin{array}{lll}
\Lambda_0 = 3 & \rho_0 = (1,1,1) & m(\Lambda_0) = 1 \\
\Lambda_2 = 1 & \rho_1 = (1,\frac{1}{3},-\frac{1}{3}) & m(\Lambda_1) = 5 \\
\Lambda_3 = -2 & \rho_2 = (1,-\frac{2}{3},\frac{1}{6}) & m(\Lambda_2) = 4
\end{array}
$$

Note that the multiplicities $m(\Lambda_i)$ of the eigenvalues of A are 1, 5, and 4, while the the symmetry classes contain 1, 3, 6 vertices, respectively.

The odd graphs $O(n)$ are distance transitive, with degree n and diameter $n-1$; its complete automorphism group of is

$$\mathbf{Aut}[O(n)] \cong S_{2n-1},$$

see [46, 17d]. The spectrum of is also known for all n

$$\Lambda_k = (-1)^k(n-k) \quad \text{with} \quad m(\Lambda_k) = \binom{2n-1}{k} - \binom{2n-1}{k-1}$$

where $0 \le k \le n-1$ indexes the distinct eigenvalues.

Cayley Graphs of the Symmetric Group. Let S_n denote, as usual, the symmetric group on n elements, and let $\mathcal{T}$ be the set of all transpositions (i,j). It is well known that the transpositions generate the symmetric group. The corresponding Cayley-graph $\Gamma(S_n, \mathcal{T})$ is hence a possible choice for the configuration space of a n-city travelling salesman problem. Little is known in general on these graphs. We will use them here in order to explain the definitions in the above sections on a non-trivial example. $\Gamma(S_n, \mathcal{T})$ has $N = n!$ vertices and is regular with degree $n(n-1)/2$. Furthermore they are bipartite since transpositions change the sign of a permutation.

The cases $n = 1$ and $n = 2$ are trivial: $\Gamma(S_1, \mathcal{T})$ is an isolated vertex, and $\Gamma(S_2, \mathcal{T})$ is the connected graph with $N = 2$ vertices.

The first case of interest is $\Gamma(S_3, \mathcal{T})$, see figure 5. It has $N = 6$ vertices, and its diameter is 2. It coincides with the *complete bipartite graph* $K_{3,3}$, which is known to be distance transitive. Therefore the symmetry classes are the distance classes. $K_{3,3}$ may serve as an example for the fact that Cayley graphs of noncommutative groups can coincide with Cayley graphs of commutative groups. Consider the commutative group with $N = 6$ elements, namely $C_2 \times C_3$, with the notation $C_2 = \{e, a\}$ and $C_3 = \{e, b, \bar{b}\}$. It is easy to check that $\Phi = \{a, ab, a\bar{b}\}$ generates $C_2 \times C_3$, and that

$$\Gamma(C_2 \times C_3, \Phi) = K_{3,3} = \Phi(S_3, \mathcal{T}).$$

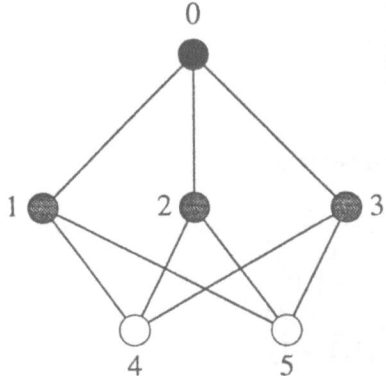

Fig. 5. The Cayley graph $\Gamma(S_3, \mathcal{T})$ coincides with the distance transitive graph $K_{3,3}$, which is also obtained as $\Gamma(C_2 \times C_3, \{a, ab, a\bar{b}\})$, where $C_2 = \{e, a\}$ and $C_3 = \{e, b, \bar{b}\}$ are the cyclic groups with 2 and 3 elements, respectively.

We emphasize at this point that the results presented in this paper depend only on the structure of the configuration space and not on the particular group representation used to construct this graph. For instance we may regard a landscape on $\Gamma(S_3, \mathcal{T})$ as a landscape on $\Gamma(C_2 \times C_3, \Phi)$, i.e., as a landscape on a CGCG, even though the physical model might force us to use the symmetric group.

The adjacency matrix and the collapsed adjacency matrix of $\Gamma(S_3, \mathcal{T})$ are

$$A = \begin{pmatrix} 0\,1\,1\,1\,0\,0 \\ 1\,0\,0\,0\,1\,1 \\ 1\,0\,0\,0\,1\,1 \\ 1\,0\,0\,0\,1\,1 \\ 0\,1\,1\,1\,0\,0 \\ 0\,1\,1\,1\,0\,0 \end{pmatrix} \qquad \hat{A} = \begin{pmatrix} 0\,1\,0 \\ 3\,0\,3 \\ 2\,0\,2 \end{pmatrix}$$

The eigenvalues, the corresponding left eigenvectors of $\hat{A}$, and the multiplicities of the eigenvalues in A are

$$\begin{array}{lll}
\Lambda_0 = 3 & \rho_0 = (1,1,1) & m(\Lambda_0) = 1 \\
\Lambda_1 = 0 & \rho_1 = (1,0,-\frac{1}{2}) & m(\Lambda_1) = 4 \\
\Lambda_2 = -3 & \rho_2 = (1,-1,1) & m(\Lambda_2) = 1
\end{array}$$

The graph $\Gamma(S_4, \mathcal{T})$ has $N = 24$ vertices; it is shown in figure 6, and its adjacency matrix is given in table 2.

A close inspection of the graph shows that there are 5 symmetry classes. The collapsed adjacency matrix $\hat{A}$ of $\Gamma(S_4, \mathcal{T})$ is

$$\hat{A} = \begin{pmatrix} 0\,1\,0\,0\,0 \\ 6\,0\,2\,3\,0 \\ 0\,1\,0\,0\,2 \\ 0\,4\,0\,0\,4 \\ 0\,0\,4\,3\,0 \end{pmatrix} ,$$

which has the following eigenvalues and left eigenvectors:

$$\begin{array}{ll}
\Lambda_0 = 6 & \rho_0 = (1,1,1,1,1) \\
\Lambda_1 = 2 & \rho_1 = (1,\frac{1}{3},-\frac{1}{3},0,-\frac{1}{3}) \\
\Lambda_2 = 0 & \rho_2 = (1,0,1,-\frac{1}{2},0) \\
\Lambda_3 = -2 & \rho_3 = (1,-\frac{1}{3},-\frac{1}{3},0,\frac{1}{3}) \\
\Lambda_4 = -6 & \rho_4 = (1,-1,1,1,-1)
\end{array}$$

The eigenvalues of $\Gamma(S_n, \mathcal{T})$ can be obtained explicitly for all n. Let $\eta = [k_1^{l_1} k_2^{l_2} \ldots k_r^{l_r}] = (\eta_1, \eta_2, \ldots, \eta_m)$ be a partition of n, i.e., $\sum_{i=1}^{r} l_i k_i = \sum_{j=1}^{m} \eta_j = m$, with $k_i > k_j$ and $\eta_i \geq \eta_j$ whenever $i < j$. It is well known that the conjugacy classes of the S_n correspond to the partitions of n via the cycle-decomposition of the permutations (see, e.g., Jacobson 1985, p. 48). The conjugacy classes of S_n are contained in the symmetry classes of the graph $\Gamma(S_n, \mathcal{T})$, see [49, Lem.2]. The eigenvalues of $\Gamma(S_n, \mathcal{T})$ have been obtained from the characters of the symmetric group. One finds

$$\Lambda_\eta = \frac{1}{2} \left[n + \sum_{j=1}^{m} \eta_j(1 - 2j\eta_j) \right] ,$$

see, e.g., [80, 81, 82].

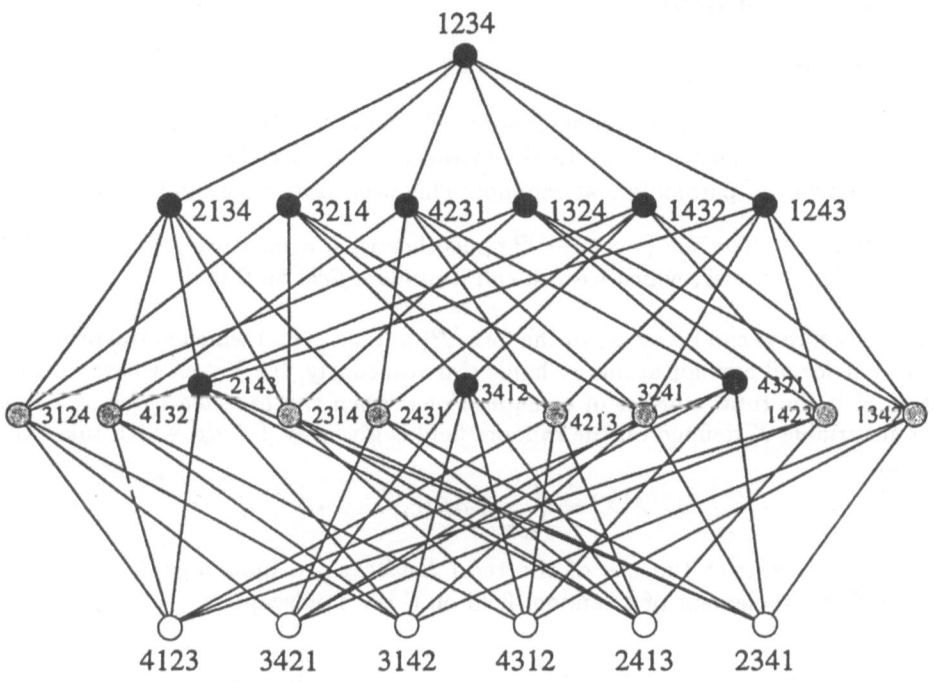

Fig. 6. The Cayley graph $\Gamma(S_4, \mathcal{T})$. Symmetry classes with respect to the reference vertex (1234) are indicated by different shadings.

A Concluding Remark. In the previous sections we have outlined a broad spectrum of tools for an analysis of the graph theoretic and algebraic properties of graphs without giving "the result". The reason for this is twofold: (i) The study of the structure of the configuration spaces is interesting in itself and provides the necessary basis for a wide range of models beyond the theory of landscapes, see for instance [75, 76]. (ii) In the investigations reported in the subsequent chapters of this contribution we will frequently require certain algebraic properties of the matrices $R^{(\mu)}$, for instance, we will need for some proofs that the $\bar{R}^{(\mu)}$ form an association scheme, and for other results we will have to require that all eigenvalues of the collapsed adjacency matrix are simple. Let me just assure the reader that all of the theory of landscapes discussed below works just fine on the most important type of configuration spaces, the Hamming graphs. On the other hand, most of it is still under construction for the more irregular classes of graphs, in particular for the Cayley graphs of the symmetric group.

3 Random Fields

3.1 Preliminaries

Definition. Many of the model landscapes mentioned in the introduction contain a stochastic element: a particular instance is generated by assigning a usually large number of parameters at random. This procedure suggests the following

Definition 1. The set $\{f : C \to \mathbb{R}\}$ together with a measure $\mu\{f\}$ forms the probability space Ξ, which we will call a *random field* on C [83].

The measure μ can be recast in the form $P(c_1, c_2, \ldots, c_N)$ which is the probability that for all configurations x_i holds simultaneously $f(x_i) < c_i$, where $c_i \in \mathbb{R}$, and N is the total number of configurations. Then the *expected value* of a random variable X defined on the random field is given by $\int X \, d\mu$ which takes the form

$$\mathcal{E}[X] = \int_{\mathbb{R}^N} X \, dP(c_1, \ldots, c_N) \, .$$

We will restrict the use of the term *landscape* to mappings $f : C \to \mathbb{R}$. Therefore, an element of a random field on C is a landscape.

Gaussian Random Fields. The well-known Gaussian probability density in $\mathbb{R}$ is

$$g_{m,\sigma^2}(x) \stackrel{\text{def}}{=} \frac{1}{\sqrt{2\pi}\,\sigma} \exp\left(-\frac{1}{2\sigma^2}(x - m)^2\right).$$

The corresponding measure is $\nu(m, \sigma^2) \stackrel{\text{def}}{=} g_{m,\sigma^2}\mu_L$, where μ_L denotes the Lebuesgue measure. The degenerate case $\nu(m, 0)$ is by definition the Dirac measure centered at m. $\nu(m, \sigma^2)$ is called a Gaussian measure for all m and σ. In the N-dimensional case one defines a measure μ to be Gaussian if for any linear map $h : \mathbb{R}^N \to \mathbb{R}$ the measure $h(\mu)$ is Gaussian. It is uniquely defined by the vector $\mathbf{m}$ of means and the nonnegative definite covariance matrix C. We will say a random field is Gaussian if its probability measure is Gaussian, $\nu(\mathbf{m}, C)$. If C is positive definite there is a probability density function

$$g_{\mathbf{m},C}(x) \stackrel{\text{def}}{=} \frac{1}{\sqrt{2\pi}^N} \frac{1}{\sqrt{\det C}} \exp\left(-\frac{1}{2}(x - \mathbf{m})C^{-1}(x - \mathbf{m})\right)$$

such that $\nu(\mathbf{m}, C) = g_{\mathbf{m},C}\mu_L$.

Averages. In many cases of practical importance, however, we are not given the random field, instead we have only a single instance (i.e., a landscape) and little or no clue from what kind of a random field it might have been drawn. This is the case, e.g., for the landscapes of RNA and protein folding, viral adaptation, and for some technical optimization problems such as the "low autocorrelated string problem" [84]. In this situation we have to be content with averages over values taken from our single instance. We will use the following notation: $\mathcal{E}[.]$ denotes averages over random fields, while angular brackets $\langle\,\rangle$ denote averages over configurations in a landscape[4]. For example we have

$$\text{Var}[f(x)] = \mathcal{E}[f(x)^2] - \mathcal{E}[f(x)]^2 \quad \text{and} \quad \sigma^2\text{var}[f] = \langle f^2\rangle - \langle f\rangle^2,$$

where the first average is taken over all sample functions of the random field f evaluated at a *fixed position* $x \in C$, while the second average runs over all vertices x in a *fixed instance* f. The averages over configurations in a single landscape are of course merely well defined finite sums. In practice, however, one has to resort to Monte-Carlo sampling techniques in order to evaluate these mean values for a problem of practical interest. In particular we have

$$\langle f\rangle \overset{\text{def}}{=} \frac{1}{N}\sum_{x\in V} f(x),$$

which gives a precise meaning to the angular brackets.

Of course the two variances $\text{Var}[f(x)]$ and σ^2 are not the same in general. This can be seen, for instance, from the following (trivial)

Example 1. Consider a (degenerate) Gaussian random field with $\mathcal{E}[f(x)] = 0$ for all x and $\mathcal{E}[f(x)f(y)] = 1$ for all x and y. A sample function drawn from this random field is almost always constant since the correlation of any two points is 1. The constant value $\bar{f}$ of this flat landscape, however, is drawn from a Gaussian distribution with mean zero and unit variance. In fact: $\text{Var}[f(x)] = 1$ for all x, while the empirical variance measured on a sample function is almost always $\sigma^2 = 0$ since almost all sample landscapes drawn from this random field are completely flat.

If, for a given quantity X, both averages give the same result in a particular model (at least in the limit of infinite system size), one says that X is a *self-averaging* quantity [2, 3].

3.2 Karhunen-Loève Decomposition

Recall that a random field on Γ is a probability space $\varXi = (\{f : C \to \mathbb{R}\}, \mu)$. An important characteristic of $\varXi$ is its covariance matrix $C = (C_{xy})_{x,y\in V}$, which is defined element-wise by

$$C_{xy} = \mathcal{E}[f(x)f(y)] - \mathcal{E}[f(x)]\mathcal{E}[f(y)].$$

---

[4] This notation should not be confused with $\langle\,,\rangle$ or $\langle\,;\rangle$ which are used for scalar products.

Denote by $\{\phi_s\}$ a complete set of orthonormal eigenvectors (which exists by symmetry of C). Although $\{\phi_s\}$ can be chosen real, we will admit complex vectors as well. Then the random field f on Γ can be represented in mean square sense as

$$f(x) \doteq \sum_{y \in \Gamma} a_y \phi_y(x), \quad \text{i.e.,} \quad \mathcal{E}\left[\left(f(x) - \sum_{y \in \Gamma} a_y \phi_y(x)\right)^2\right] = 0.$$

This series representation is known as the *Karhunen-Loève* decomposition of the random field Ξ. It coincides for finite sets with the well known *principal component analysis* introduced by Hotelling in 1933 [85], see also [86]. Its most important property is

Theorem 2. *[85] The coefficients a_y are uncorrelated with respect to the measure μ, i.e., $\mathcal{E}[a_x a_y^*] = \mathcal{E}[a_x]\mathcal{E}[a_y^*]$ if $x \neq y$.*

3.3 The Markov Property

Properties related to the Markov property of time series have been first studied in the context of spatial stochastic processes. There are two non-equivalent definitions of nearest neighbor models in the literature, one due to Bartlett [87, sect. 2.2] based on conditional probabilities, and the other one due to Whittle [88] based on a particular product form of the joint probability distribution $P(\mathbf{c})$. These two approaches have been linked in the early seventies by the Hammersley-Clifford theorem, which we will discuss below.

Let $p(c_1, c_2, \ldots, c_N)$ be a probability density function (or, if the number of possible states c_k is countable, the probability for the $f(x_i) = c_i$ for all i). Recall that in the continuous case we have

$$p(c_1, c_2, \ldots, c_N) = \frac{\partial^N}{\partial c_1 \partial c_2 \ldots \partial c_n} P(c_1, c_2, \ldots, c_N)$$

in terms of the cummulative probabilities used above.

Definition 3. A random field Ξ on a graph Γ has the *Markov Property* if the conditional probability density

$$p(c_x | c_1, c_2, \ldots c_{x-1}, c_{x+1}, \ldots c_N)$$

depends for each $x \in V$ only on the neighbors of x, that is, on the values c_y of all all vertices $y \in N(x)$.

Theorem 4.   *(Hammersley-Clifford)*
Let $p(\mathbf{c}) > 0$ be an absolutely continuous probability density function (in the continuous case, or the probabilities of the states $\mathbf{c}$ in the discrete case). Then

the following holds:
The random field Ξ is Markovian if and only if there is a function

$$Q(\mathbf{c}) \stackrel{\text{def}}{=} g_0 + \sum_{i=1}^{N} c_i g_i(c_i) + \sum_{i<j} c_i c_j g_{ij}(c_i, c_j) + \sum_{i<j<k} c_i c_j c_k g_{ijk}(c_i, c_j, c_k) + \dots$$

fulfilling $p(\mathbf{c}) = \exp(Q(\mathbf{c}))/ \int \exp(Q(\mathbf{c}))\,\mathrm{d}\mathbf{c}$ such that $g_{i_1, i_2, \dots, i_p}$ vanishes identically whenever the subgraph of Γ induced by vertex set $\{x_{i_1}, x_{i_2}, \dots, x_{i_p}\}$ is not complete, i.e., if not all pairs of these vertices are edges in Γ.

Proof. The original Proof by Hammersley and Clifford (1971) has never been published. A detailed discussion can be found in Besag's paper [83]. A proof for discrete variables can be found in [89]. Moussouris [90] gives three different proofs and shows that the positivity condition cannot be omitted. Less general results for regular lattices are described in [91, 92, 93].   □

The denominator $\int \exp(Q(\mathbf{c}))\,\mathrm{d}\mathbf{c}$ is reminiscent of the partition function in statistical physics. A very detailed theory is available for stationary Gaussian Markov time series. It remains to be seen if a comparably rich and interesting structure can be developed also for the class of Markovian random fields on graphs. A simple result in this direction will be discussed at the end of the following section.

3.4 Isotropy

Characterization of Isotropic Random Fields.

Definition 5. A random field on a graph Γ is *isotropic* if and only if

(i) $\mathcal{E}[f(x)] = a_0$ for all configurations $x \in \Gamma$.
(ii) Suppose the pairs of configurations (x, y), (u, v) belong to the same symmetry class $\bar{\mu}$.
    Then $\mathcal{E}[f(x)f(y)] = \mathcal{E}[f(u)f(v)]$.

Remark. The notion of isotropy for random fields is analogous to the defintion of stationarity for stochastic processes. Following the conventions of Karlin and Taylor [94] our notion of isotropy should be called "covariance isotropic", "weakly isotropic", or "wide sense isotropic". Weinberger's [59] definition corresponds to "strictly isotropic". For a Gaussian random field the notions of (weak) isotropy and strict isotropy coincide.

Lemma 6. *Let Γ be a vertex transitive graph. Then a random field on Γ is isotropic if and only if*

(i) $\mathcal{E}[f(x)] = a_0$ for all $x \in V$, and

(ii) there exist a constant $C_{00} = Var[f(x)]$ and a real valued function ρ of the symmetry classes with $\rho(0) = 1$ such that

$$C = C_{00} \sum_{\mu} \rho(\mu) R^{(\mu)}.$$

Proof. This is a simple rewriting of the above definition, observing that all entries in the covariance matrix belonging to a fixed symmetry class μ are the same and correspond to the non-zero entries of $R^{(\mu)}$.   □

Note that by symmetry of the covariance matrix C we have $\rho(\mu) = \rho(\mu^+)$, i.e., $C = C_{00} \sum_{\tilde{\mu}} \rho(\tilde{\mu}) R^{(\tilde{\mu})}$. The function $\rho(\mu)$ is called the *autocorrelation function* of the random field. It will be discussed in section 3.3.

Theorem 7. *Let Γ be a vertex transitive graph for which the matrices $R^{(\tilde{\mu})}$ form a symmetric association scheme, and let Ξ be an isotropic random field on Γ. Then the adjacency matrix A of Γ and the covariance matrix C of the random field f commute, i.e., Fourier decomposition and Karhunen-Loève decomposition coincide.*

Proof. By assumption there is a commutative algebra $\langle \mathfrak{R} \rangle$ spanned by the matrices $R^{(\tilde{\mu})}$. Since C is a linear combination of these matrices we have $C \in \langle \mathfrak{R} \rangle$. Of course A is contained in this algebra as well, and hence $AC = CA$.   □

Remark. The case of CGCG is dealt with in [59]. The case of distance transitive graphs is proved in [95], and an extension to vertex transitive graphs for which the commuting algebra coincides with the adjacency algebra is discussed in [49].

As a consequence we immediately conclude that the Fourier coefficients are uncorrelated for an isotropic random field on a vertex transitive graph with the Θ-property, see sect. 2.5.3. A more detailed result can be obtained for isotropic landscapes on a CGCG.

Theorem 8. *A random field on a CGCG is isotropic if and only if*

(i) $\mathcal{E}[a_l] = 0$ for all $l > 0$; the coefficient a_0 is arbitrary.
(ii) The coefficients are pairwise uncorrelated, $\mathcal{E}[a_k a_l] = \mathcal{E}[|a_l|^2]\delta_{kl}$.
(iii) $\mathcal{E}[|a_l|^2] = V(\mu)$ depends only on the symmetry class μ to which l belongs.

Proof. A complete proof is given in [49] based on [59].   □

Note that this theorem depends on the fact the we can relate the eigenfunctions $e_p(x)$ directly with symmetry classes. This is not possible, however, for more general graphs [49]. The relation of Fourier coefficients and isotropy on more general graphs certainly deserves attention in forthcoming investigations.

Landscapes on the Hypercube and the p-Spin Models. The vertices of a Boolean hypercube are the sequences of length n taken from an alphabet of size $\alpha = 2$. Without losing generality we may use the alphabet $\{+1, -1\}$. A configuration is then a string σ of "spins" $\sigma_k \in \{+1, -1\}$. An alternative encoding uses a binary string x, with $x_i \in \{0, 1\}$. The following result is well known in the literature:

Lemma 9. *Any landscape f on the Boolean Hypercube can be written as*

$$f(\sigma) = J_0 + \sum_{p=1}^{n} \sum_{i_1 < i_2 < \ldots < i_p} J_{i_1 i_2 \ldots i_p} \sigma_{i_1} \sigma_{i_2} \cdots \sigma_{i_p} ,$$

where the $J_{i_1 i_2 \ldots i_p}$ are constants.

Proof. See, e.g., [49, 59]. $\qquad\qquad\qquad\qquad\qquad\qquad\qquad\qquad\qquad\qquad\qquad$ □

The p-spin Hamiltonian

$$\mathcal{H}_p(\sigma) = \sum_{i_1 < i_2 < \ldots < i_p} J_{i_1 i_2 \ldots i_p} \sigma_{i_1} \sigma_{i_2} \cdots \sigma_{i_p}$$

was introduced by Derrida [96] in order to bridge the gap between the SK model [97], which is the special case $p = 2$ and the random energy model [96, 98, 99, 100]. The coefficients $J_{i_1 i_2 \ldots i_p}$ are i.i.d. Gaussian random variables with mean 0 and unit variance. [5] The autocorrelation function of the p-spin Hamiltonian $\mathcal{H}_p(\sigma)$ is simply $\rho = \rho_p$ for all p [21]. Finally we remark that $\varepsilon_q(\sigma) \overset{\text{def}}{=} \sigma_{i_1} \sigma_{i_2} \ldots \sigma_{i_p}$ is of course only an alternative representation of the Fourier basis vectors $e_q(x)$ on the Boolean hypercube.

Autocorrelation Functions of Isotropic Random Fields. As usual let Γ be a vertex transitive graph. Consider a simple random walk $\{x_0, x_1, \ldots\}$ on Γ, see sect. 2.3.1. Weinberger [101] has suggested to use the time series obtained by evaluating the random field at each step in order to gather information on the random field itself. The correlation function along such a time series is defined by

$$r(s) = \frac{\mathcal{E}[f(x_{t+s}) f(x_t)] - \mathcal{E}[f(x_{t+s}) f(x_t)]}{\sqrt{\text{Var}[f(x_{t+s})] \text{Var}[f(x_t)]}} .$$

For isotropic random fields this reduces to

$$r(s) = \frac{\mathcal{E}[f(x_s) f(x_0)] - \mathcal{E}[f(x_0)]^2}{\text{Var}[f(x_0)]} = \sum_{\mu} \varphi_{s\mu} \rho(\mu) .$$

In the light of this equation it is not surprising that there is a simple relation between the autocorrelation functions $\rho(\mu)$ and $r(s)$.

---

[5] In Derrida's original formulation the variance was chosen to be $S_p^2 = p!/n^p$, such that the variance of the Hamiltonian itself is independent of n.

Theorem 10. *Let Γ be a vertex transitive graph and consider an isotropic random field on Γ with correlation functions $\rho(\mu)$ and $r(s)$, respectively. Then:*

(i) $r(s) = \lambda^s$ if and only if $\rho \hat{A} = \Lambda \rho$.
(ii) Let $\{\rho_p\}$ be the orthonormal system of left eigenvectors of $\hat{A}$ described in section 2.7. We have $r(s) = \sum_p b_p \lambda_p$ if and only if $\rho(\mu) = \sum_p b_p \rho_p$. Of course, such a decomposition always exists.

Proof. This theorem is the content of [57].   □

An immediate consequence is the following surprising relation between eigenvalues and eigenvectors of a large class of graphs.

Corollary 11. *Let Γ be symmetric, i.e., assume that Γ is vertex transitive and that it has a symmetry class*

$$\ddot{\mathbf{1}} = \{ (x, y) \mid d(x, y) = 1 \}$$

consisting of all pairs of neighbors. Then $\rho_p(1) = \lambda_p$.

Proof. See [49].   □

Gaussian Markovian Isotropic Random Fields. A stationary Gaussian Markov process in $\mathbb{R}$, a so-called Ornstein-Uhlenbeck process or AR(1) process, has an autocorrelation function of the form $\rho(t) = \exp(-at)$. Gaussian Markovian isotropic random fields present an immediate generalization. Hence we are interested in the form of their covariance matrices. For simplicity we will assume that the Gaussian distribution is non-degenerate (i.e., that the covariance matrix C is invertible) and that the configuration space Γ is a symmetric graph.

Theorem 12. *Let Γ be symmetric and let Ξ be a Markovian Gaussian isotropic random field on Γ. Then its covariance matrix is of the form*

$$C = a_0 (I + \alpha A)^{-1}$$

Proof. Consider the function $Q(\mathbf{c})$. Since Ξ is Gaussian we have

$$Q(\mathbf{c}) = -\frac{1}{2} \left((\mathbf{c} - \mathbf{m}) C^{-1} (\mathbf{c} - \mathbf{m}) \right)$$

where $\mathbf{m}$ is the vector of expected values $m_x \stackrel{\text{def}}{=} \mathcal{E}[f(x)]$. Isotropy implies $\mathbf{m} = m_0 \mathbf{1}$. Without losing generality we can assume therefore that $m_0 = 0$. The Markov property now requires that $(C^{-1})_{xy} = 0$ unless $x = y$ or x and y are neighbors in Γ. Since C is an element of the commuting algebra of Γ so is C^{-1}, i.e., C^{-1} must be a linear combination of the matrices $R^{(\mu)}$. Symmetry of Γ implies that the identity I and $R^{(1)} = A$ are the only $R^{(\mu)}$ matrices which have non-zero entries only in the diagonal and for pairs of neighbors. Thus the covariance matrix must be of the form $C^{-1} = b_0 I + b_1 A$.   □

3.5 Homogeneity

The definition of isotropy becomes void if the graph Γ does not have symmetries. In fact, almost all graphs have only the trivial group of automorphisms. Nevertheless, it is possible to consider random fields and their correlation structure on such objects. To this end we introduce the following

Definition 13. A random field Ξ on a graph Γ is *homogeneous* if

(i) $\mathcal{E}[f(x)] = a_0$ for all $x \in V$, and
(ii) $\mathcal{E}[f(x)f(y)] = \mathcal{E}[f(u)f(v)]$ whenever $d(u, v) = d(x, y)$.

Note that isotropy and homogeneity are the same thing if Γ is distance transitive.

Homogeneous random fields play a special role on distance regular graphs. Isotropy is in general trivial on this class of graphs, since distance regular graphs need not even be vertex transitive. In particular, the main result of sect. 3.4.3 holds for homogeneous random fields on distance regular graphs. In general, however, isotropy proves to be the more powerful concept for deriving interesting theorems. In particular, the condition for obtaining exponential autocorrelation functions along simple random walks on a non distance transitive graph requires in general different covariances for pairs of vertices in different symmetry classes which may well belong to the same distance class. Thus there will in general be no homogeneous random fields with this property.

3.6 Superpositions of Random Fields

Definition. Note that we regard a p-spin model as random field on the Boolean Hypercube. For the next result we have to state precisely what we mean by a "sum" of random fields.

Definition 14. Let $\Xi_i = (\{f : V \to \mathbb{R}\}, \mu_i)$ be random fields on Γ. The *superposition* of two random fields on Γ with fixed coefficients $a, b \in \mathbb{R}$ is the probability space

$$a\Xi_1 + b\Xi_2 \stackrel{\text{def}}{=} (\{f : V \to \mathbb{R}\}, \mu^*) \quad \text{with} \quad \mu^*(f) = \int_g \mu_1(\frac{1}{a}g)\mu_2(\frac{1}{b}(f - g)).$$

The effect of this definition is that elements of the probability spaces Ξ_i are added as independent random variables.

Theorem 15. *A Gaussian isotropic random field on the Boolean hypercube is a superposition of p-spin models with independent coefficients.*

Proof. See [49]. □

The Random Energy Model. The results derived in the previous section allow us to calculate a "spectrum" $\Omega(\ell) = \Omega_\ell$ of the autocorrelation functions: $\rho(d) = \sum_{\ell=0}^{n} \Omega_\ell \rho_\ell(d)$.

We consider here a random field on a vertex transitive graph, where the values $f(x)$ are i.i.d. random variables (see, e.g., [102]. This Random Energy Model (REM) introduced by Derrida [96, 98, 99] has been applied to a variety of problems, in particular to the maturation of the immune system [18, 14].

Lemma 16. *Let Γ be vertex transitive. The REM is isotropic and has autocorrelation function $\rho(\mu) = 0$ for all $\mu \neq 0$.*

Proof. See [49].   □

As a consequence a Gaussian REM on a Boolean hypercube – that is, Derrida's original model – is a superposition of p-spin models. Derrida argued that the REM is the limit $p \to \infty$ of the p-spin models. The following theorem makes the sense in which this is true more precise.

Theorem 17. *Let Γ be vertex transitive. Then the spectrum of the REM is*

$$\Omega(\ell) = \frac{1}{\langle \rho_\ell; \rho_\ell \rangle}$$

On a Boolean hypercube $\mathbf{S}_2^n$ we have explicitly: $\Omega(p) = \frac{1}{2^n}\binom{n}{p}$.

Proof. See [49].   □

Hence the REM on the Boolean hypercube can be represented as a superposition of a "continuum" of p-spin models with $p \approx n/2$.

Nk Models. The Nk model assigns a real valued "fitness" to the bit string **b** by first assigning a real valued "fitness contribution", f_i , to the ith bit, b_i, in **b**. Each such assignment depends, not just on i and the value of b_i, but also on $0 \leq k < n$ other bits, which we call its "neighbors". The fitness contribution of each site is a random function, $f_i(\mathbf{b}_i)$, of the substring, $\mathbf{b}_i$, formed by the ith bit and its k neighbors. $f_i(\mathbf{b}_i)$ is assigned by selecting an independent random variable from some distribution, such as the uniform or Gaussian distributions, for each of the 2^{k+1} possible values of $\mathbf{b}_i$, thus generating a "fitness table" for the ith site. There is a different, independently generated table for each of the n sites. Then, given any string of n bits, the total fitness of the string, f, is defined as the average of the fitness contributions of each site; that is,

$$f(\mathbf{b}) = \frac{1}{n}\sum_{i=1}^{n} f_i(\mathbf{b}_i).$$

The Nk model comes in different "flavors" depending on the choice of neighbors influencing a particular site fitness. The simplest — but not the only —

way of choosing neighbors, at least for even k, is to use the k sites adjacent to site i; that is, the bits at sites $i - k/2$ through $i + k/2$. As in the original formulation of the model, we introduce periodic boundary conditions to assign neighbors to sites i with $i \leq k/2$ and $i \geq n - k/2$. In other words, we assume that the sites are arranged in a circle, such that site n is next to site 1. (Periodic boundary conditions are chosen because they minimize chain length dependent end effects and because we are interested in bulk properties only). This choice of neighbors gives rise to a class of short range spin glasses. We will call this arrangement the "adjacent neighborhood" model, AN. Alternatively, we could assign the neighbors by randomly selecting, for each site i, k other sites. This assignment of neighbors makes the model similar to a long range, dilute spin glass. It will be referred to as "random neighborhood" model, RN. In a third variation we drop the requirement that f_i depends on bit i. This version will be called "purely random" model, PR. The Nk model reduces to the REM for $k + 1 = n$, see e.g., [103].

Lemma 18. *The autocorrelation functions of the Nk model with random, purely random, and adjacent neighborhood, respectively are given by*

$$\rho_{RN}(d) = (1 - \frac{d}{n})(1 - \frac{k}{n-1})^d$$

$$\rho_{PR}(d) = (1 - \frac{k+1}{n})^d$$

$$\rho_{AN}(d) = 1 - \frac{k+1}{n}d + \frac{1}{\binom{n}{d}} \sum_{j=1}^{\min(k,n+1-d)} (k - j + 1)\binom{n-j-1}{d-2}.$$

Proof. For a proof see, e.g., the appendix in [20].   □

Lemma 19. *The autocorrelation function of the adjacent neighbor Nk model, ρ_{AN}, is a polynomial of degree $k + 1$. Its autocorrelation function is therefore of the form*

$$\rho_{AN}(d) = \sum_{p=0}^{k+1} \Omega_p \rho_p(d).$$

Proof. See [49, Thm.11].   □

The coefficients $\Omega(\ell)$ can be obtained explicitly for sequence spaces, if the correlation function $\rho(d)$ is known.

$$\Omega(\ell) = \frac{\langle \rho_\ell; \rho \rangle}{\langle \rho_\ell; \rho_\ell \rangle} = \frac{n!}{\alpha^n} \sum_{i=0}^{n} \rho(i) \sum_{j=0}^{p} \frac{(-1)^j (\kappa - 1)^{p+i-j}}{j!(i-j)!(p-j)!(n+j-i-p)!}$$

For Nk models this expression has been evaluated numerically, results are shown in figure 7.

Weinberger and Stadler [21] concluded based on a comparison of correlation lengths, that Nk models roughly resemble p-spin models with $p = (k+1)/2$. This

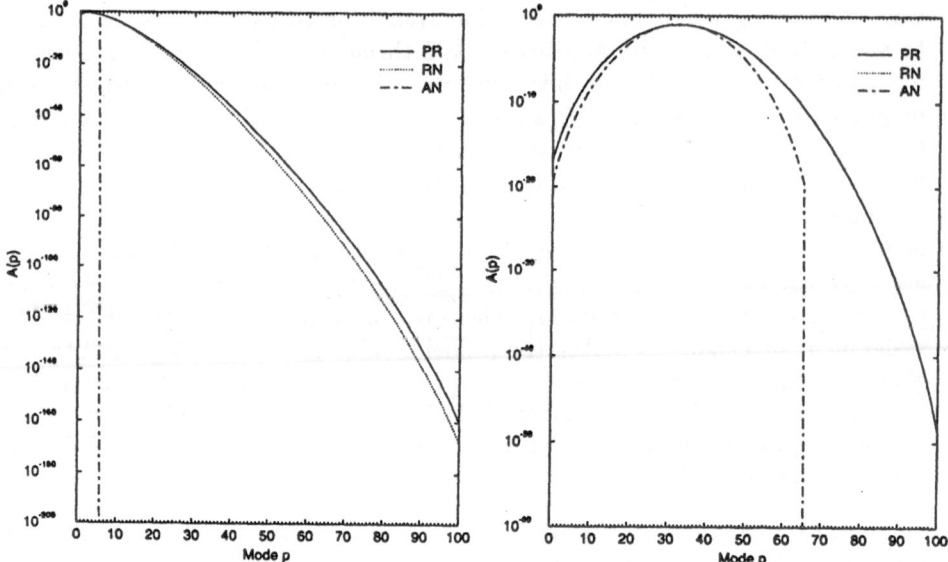

Fig. 7. Spectral decompositions of Nk models on Boolean hypercubes with $n = 100$. The logarithmic plots emphasize that the two random neighborhood Nk models involve all modes $0 \le p \le n$ for all k while the AN models are superpositions of the modes $0 \le p \le k + 1 \le n$.

view is supported by the spectra of the Nk models showing that the dominating modes are those centered around $\bar{\pi} = (k + 1)/2$ for all three versions of the Nk model [49]. The convergence of Nk models and p-spin models is similar to the case of the p-spin models and the REM. If k scales linearly with n, i.e., if $(k + 1)/n \to \chi > 0$, then we expect that the spectrum in terms of the scaled variables $\psi = p/n$ will be concentrated at $\bar{\psi} = \chi/2$ in the limit $n \to \infty$.

Superposition of Nk Models. We have seen that all isotropic random fields on the Boolean hypercube can be constructed by superpositions of p-spin models. Kauffman's Nk-models cannot be used for this purpose. Let $\{\mathcal{N}_k(x)\}$ denote a set of Nk models with parameter k, such that

$$\mathcal{E}[\mathcal{N}_i(x)\mathcal{N}_j(y)] - \mathcal{E}[\mathcal{N}_i(x)]\mathcal{E}[\mathcal{N}_j(y)] = 0$$

whenever $i \ne j$, i.e., we suppose that the entries in the fitness tables are uncorrelated also between the random fields. Given the construction of the Nk model this is the natural assumption. Then we have the following

Theorem 20. *Let Ξ be an isotropic random field of the form $\Xi = \sum_{k=0}^{k_{max}} \beta_k \mathcal{N}_k$, i.e., Ξ is a superposition of Nk models. Then $\Omega(p) > 0$ for all $p \le k_{max} + 1$. Furthermore $\Omega(p) > 0$ for all $p \le n$ if at least one of the Nk models $\mathcal{N}_k$ is an RN or a PR version of the Nk model.*

Proof. The result is obtained by a direct calculation of the covariance matrix, see [49]. □

Corollary 21. *A p-spin model cannot be represented as a superposition of Nk models for all $p \geq 1$. The empirical correlation functions of the 1-spin model and the Nk-model with $k = 0$ coincide. The empirical correlation function of a p-spin model with $p \geq 2$ cannot be modeled by a superposition of Nk models.*

3.7 Transformations

Operators on Landscapes and Random Fields. Let $f : C \rightarrow \mathbb{R}$ be a landscape on C and let $F : R^{|V|} \rightarrow R^{|V|}$ be a vector field. Then we explain $F[f] : C \rightarrow \mathbb{R}$ component-wise as

$$F[f](x) \stackrel{\text{def}}{=} (F(f(x_1), f(x_2), \ldots, f(x_N)))_x ,$$

i.e., we interpret F as an operator acting on the landscape f. Before we proceed we briefly introduce the two most common examples:

(i) The averaging (or smoothing) operator Ψ is defined in the above notation as

$$\Psi[f](x) \stackrel{\text{def}}{=} \frac{1}{1+N(x)} \left(f(x) + \sum_{y \in N(x)} f(y) \right) = \frac{1}{1+N(x)}[(I+A)f]_x.$$

Ψ is of course a linear operator, its effect is averaging the fitness values over a ball of radius 1. Consequently it amounts to applying the matrix $(I+D)^{-1}(I+A)$ to the vector of fitness values.

(ii) Let $\beta : \mathbb{R} \rightarrow \mathbb{R}$ be an arbitrary function. Then $\beta[f]$ is explained as the composition of f and β, i.e.,

$$\beta[f](x) \stackrel{\text{def}}{=} \beta(f(x)).$$

Now let $\Xi = (\{f : C \rightarrow \mathbb{R}\}, \mu)$ be a random field on C. Then

$$F[\Xi] \stackrel{\text{def}}{=} (\{F[f] : C \rightarrow \mathbb{R}\}, \mu)$$

is again a random field on C. In the following we will briefly discuss some properties of the random fields $F[\Xi]$ which are obtained from a simple random field Ξ by means of the two types of operators described above.

Averaging and Iterated Smoothing Landscapes. The basic properties of linear transformations of random fields are compiled in the following

Theorem 22. *Let C be the covariance matrix of a random field Ξ and let Ω be a linear symmetric operator.*

(i) The covariance matrix of $\Upsilon = \Omega[\Xi]$ is given by $C^\Omega = \Omega C \Omega$.

(ii) If Ξ is isotropic and Γ is a vertex transitive graph for which the matrices $\hat{R}^{(\mu)}$ form a symmetric association scheme, then $C^\Omega = (\Omega^2)C$.

(iii) If Γ is a in (ii) and $\Omega \in \mathfrak{A}[\Gamma]$ (that is, Ω is a polynomial in A), then $\Upsilon = \Omega[\Xi]$ is again isotropic on Γ and $\rho^\Omega = c \cdot \hat{\Omega}^2$, where the constant $c = 1/(\hat{\Omega}^2 \rho)(0)$ is chosen such that $\rho^\Omega(0) = 1$.

(iv) Under the assumptions of (iii) holds $\rho^\Omega(\mu) = c \cdot \sum_\nu a_\nu P^2(\lambda_\nu)\rho_\nu(\mu)$, where λ_ν denotes the eigenvalue of $\hat{A}$ belonging to ρ_ν. In particular, if $\rho = \rho_j$ for some j, then we have $\rho^\Omega = \rho$, i.e., eigenfunctions of $\hat{A}$ remain unchanged.

Proof. See [05]. □

Let us now consider the smoothing operator $\Psi \overset{\text{def}}{=} \frac{1}{D+1}(A + E)$. Suppose Γ is distance transitive. Then we obtain as an immediate consequence of the above theorem the following explicit representation of the autocorrelation function

$$\rho^\Psi(d) = \frac{\sum_{k=-2}^2 \alpha_k(d)\rho(d+k)}{\sum_{k=0}^2 \alpha_k(0)\rho(k)}$$

where $\alpha_k(d) = (\hat{\Psi}^2)_{d,d+k}$. The coefficients have a simple combinatorial interpretation: $\alpha_k(d)$ is the number of pairs (u,v) with $u \in N(x)$, $v \in N(y)$ with $d(u,v) = d+k$ where $d(x,y) = d$ is fixed.

For particular graphs Γ it is fairly easy to calculate the coefficients $\alpha_k(d)$ explicitly. On Hamming graphs, for instance, one obtains [95]

$$\alpha_{-2}(d) = d(d-1)$$
$$\alpha_{-1}(d) = 2d + (\alpha-2)(2d-1)d$$
$$\alpha_0(d) = 1 + (\alpha-1)(n + 2d(n-d)) + (\alpha-2)(\alpha d + d(d-1)(\alpha-2))$$
$$\alpha_{+1}(d) = 2(\alpha-1)(n-d) + (\alpha-2)(\alpha-1)(n-d)(2d+1)$$
$$\alpha_{+2}(d) = (\alpha-1)^2(n-d)(n-d-1)$$

Let Ξ_0 be the random energy model. The autocorrelation function is $\rho(d) = \delta_{0,d}$. The corresponding empirical correlation function is $\hat{\rho}(0) = 1$ and $\hat{\rho}(d) = -1/(N-1)$ for $d > 0$, respectively, where N denotes the number of points in the configuration space, see sect. 4.1. Define $\Xi_q = \Psi[\Xi_{q-1}]$. Let $\hat{\rho}^{(q)}$ denote the empirical autocorrelation function expected for an instance of the random field Ξ_q. We will refer to this family of landscapes as *iterated smoothing landscapes*, ISLs. The numerical data shown in figure 8 indicate that they form indeed a family of tuneably rugged random fields, just like the Nk models. In particular, we have the following

Theorem 23. *Let Γ be a generalized hypercube over an alphabet with $\alpha > 2$ letters. Then*

$$\lim_{q \to \infty} \hat{\rho}^{(r)}(d) = 1 - \frac{1}{n}\frac{\alpha}{\alpha-1}d.$$

On a Boolean Hypercube ($\alpha = 2$) we have

$$\lim_{q \to \infty} \hat{\rho}^{(r)}(d) = \frac{n}{n+1}\left(1 - \frac{2d}{n}\right) + \frac{1}{n+1}(-1)^d.$$

Proof. See [95]. $\square$

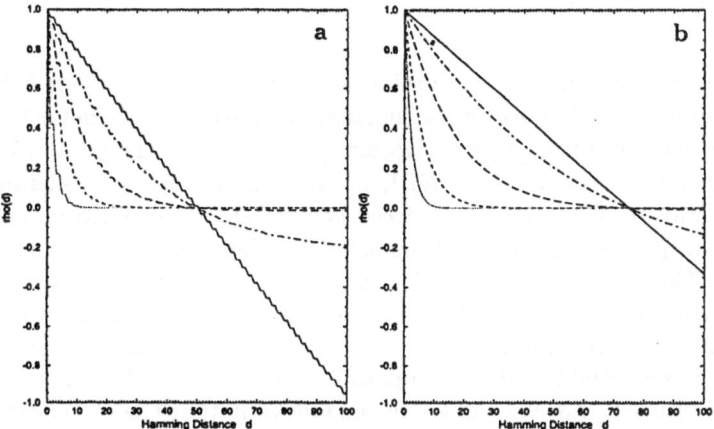

Fig. 8. Autocorrelation function of ISLs on generalized hypercubes with **a)** $\alpha=2$, and **b)** $\alpha=4$ for with $n=100$. The number of smoothings are **a)** $r=25$ (dotted), $r=50$ (short dashed), $r=75$ (long dashed), $r=100$ (dot-dashed), $r=200$ (solid), and **b)** $r=50$ (dotted), $r=100$ (short dashed), $r=150$ (long dashed), $r=200$ (dot-dashed), and $r=400$ (solid), respectively.

We remark that the contribution of $\hat{\rho}_n$ decreases with the size of the configuration space. The limiting autocorrelation function corresponds to both the trivial spin glass model with Hamiltonian $\mathcal{H}(\sigma) = \sum_j A_j \sigma_j$, and to the Nk-model with $k = 0$. We also note that for $q/n \ll 1$ the landscapes are essentially uncorrelated, i.e., the correlation length is $o(n)$. Numerical estimates show that the nearest neighbor correlation $\hat{\rho}^{(q)}(1)$ depends only on the ratio q/n for large n, see [95] for details.

Instead of starting with a REM one can as well apply the averaging operator to spin glass and Nk models. For p-spin models one finds

Theorem 24. *Let Ω be any polynomial of the adjacency matrix of the Boolean Hypercube. Then the landscapes $\mathcal{H}_p$ and $\Omega[\mathcal{H}_p]$ have the same autocorrelation functions.*

Proof. See [95, Thm.14]. $\square$

In contrast to the p-spin models, the landscapes of Nk type are affected by the smoothing operator.

Theorem 25. *Let f be an arbitrary Nk landscape. Then the autocorrelation function of the landscapes $\Psi^r[f]$ converges to $\rho(d) = \xi(1 - 2d/n) + (1 - \xi)(-1)^d$ with $0 < \xi \leq 1$ as q tends towards infinity.*

Proof. See [95, Thm.15].   □

The variants of the Nk model do not behave identically under the action of Ψ. Whenever the autocorrelation function is a polynomial of degree less than n, then we have $\xi = 1$ in the theorem above. For exact expressions of the autocorrelation functions of various Nk models see sect. 3.6.3. The theorem holds with $\xi = 1$ for all alphabets with $\alpha \geq 3$ letters.

Non-Linear Transformations. Transforming a landscape or a random field with a non-linear function is important for instance in (bio)physical chemistry, since the chemical reaction rate k and activation energies $\Delta G^{\neq}$ are related by Arrhenius' law [104]: $k = A \exp(-\Delta G^{\neq}/RT)$, where R is a universal constant and T is the temperature. Consider now a fixed type of chemical reaction, say binding to a fixed target. Then $\Delta G^{\neq}$ will be sequence dependent, and thus it forms a landscape. Consequently the reaction rate constants form a landscape as well. If we assume, for simplicity, that the pre-exponential factor A is not sequence dependent, we observe immediately that both landscapes have the same geometry since the exponential function is monotonic and thus maps local optima to local optima, see also sect 4.3. We will see in the following that the correlation structure, however, does not remain unchanged.

As an example we consider here the correlation function of an exponential transformation $F(z) = \exp(qz)$ of a Gaussian isotropic random field with covariance matrix C. It is convenient to define the correlation coefficient $\rho(x, y) \overset{\text{def}}{=} \text{var}[f]^{-1} C_{xy}$. The covariance of the transformed random field is given by

$$C_{xy}^* = \mathcal{E}[F[f(x)]F[f(x)]] - \mathcal{E}[F[f(x)]]\mathcal{E}[F[f(x)]].$$

For Gaussian random fields these expectation values are simple Gaussian integrals which can be evaluated explicitly. The details can be found in [32]. The final result is

$$\rho^*(x, y) = \frac{a^{\rho(x,y)} - 1}{a - 1},$$

where $a = e^{q^2}$ depends on the parameter q in the transformation. The function $f(t) = \frac{a^t - 1}{a - 1}$ is concave for all t and all $a > 1$, i.e. all $q \neq 0$, and furthermore $f(0) = 0$ and $f(1) = 1$. It follows immediately that $|\rho(x, y)| > |\rho^*(x, y)|$ whenever $\rho(x, y) \neq 0$, 1, or -1.

Suppose the correlation function is of the form $\rho(d) = 1 - \frac{1}{\ell}d + o(d)$, where ℓ is the correlation length, which will be discussed in sect. 4.1.4 in detail. A simple calculation then shows that $\rho^*(d)$ has the same form, with correlation length

$$\ell^* = \frac{a - 1}{a \log a} \ell \approx \frac{1}{q^2} \ell$$

for large values of q. Thus the a-dependent pre-factor becomes very small for large absolute value of q. Thus an exponential transformation can lead to a random field with very small correlation without changing the rank order of the fitness values, and thus without disturbing the distribution of local optima.

4 Landscapes

Let us now consider individual landscapes instead of random fields. We will suppose that we have a means of computing (or measuring) the value of the landscape $f : V \to \mathbb{R}$ for all configurations $x \in V$. This means that we have now a fixed matrix of interaction coefficients in spin glass model or a fixed matrix of bond lengths for a TSP. Models for biologically important landscapes will be discussed in section 5.

4.1 Empirical Correlation Functions

Characterization of Empirical Correlation Functions. The empirical autocorrelation function of a landscape on a vertex transitive graph is defined as [23, 20, 24, 29]

$$\hat{\rho}(\mu) \stackrel{\text{def}}{=} \frac{\langle f(x)f(y)\rangle_{(x,y)\in\bar\mu} - \langle f\rangle^2}{\langle f^2\rangle - \langle f\rangle^2}.$$

Of course, the autocorrelation function is invariant under the transformation $f \to f - \langle f\rangle$. Therefore we can assume without losing generality $\langle f\rangle = 0$ for the remainder of this chapter. Under this assumption we may rewrite the definition above in the form

$$\hat{\rho}(\mu) = \frac{1}{|\mu|} \frac{\langle f, R^{(\mu)}f\rangle}{\langle f, f\rangle},$$

where $\langle .\,,.\rangle$ denotes as usual the standard scalar product on Euclidean vector spaces. Of course we have $\hat{\rho}(\mu) = \hat{\rho}(\mu^+)$ in complete analogy with the correlation functions of isotropic random fields. In fact, the above definition tacitly assumes that the landscape is "isotropic" in a certain sense, since we average over entire symmetry classes.

The appropriate definition of a correlation function on a distance regular graph is analogous to homogeneous landscapes:

$$\hat{\rho}_h(d) \stackrel{\text{def}}{=} \frac{\langle f(x)f(y)\rangle_{d(x,y)=d}}{\langle f^2\rangle} = \frac{1}{|N_d|} \frac{\langle f, A^{(d)}f\rangle}{\langle f, f\rangle},$$

where we have again used $\langle f\rangle = 0$. Note that for distance transitive graphs both definitions coincide.

In the following we characterize the set of all possible empirical autocorrelation functions for sufficiently symmetric configuration spaces.

Theorem 1. *Let* $\{\rho_\mu\}$ *be an orthogonal basis of left eigenvectors of the collapsed adjacency matrix of a* CGCG Γ. *Then* $\hat{\rho}$ *is an autocorrelation function of a landscape on* Γ *if and only if*

$$\hat{\rho} = \sum_\mu \Omega_\mu \rho_\mu, \quad \Omega_0 = 0, \quad \Omega_\mu \geq 0, \quad \sum_\mu \Omega_\mu = 1.$$

The coefficients Ω_μ *can be obtained from the Fourier coefficients of* f:

$$\Omega_\kappa = \frac{\sum_{g\in\kappa} |a_g|^2}{\sum_{g\in G} |a_g|^2}.$$

Proof. See [49, Thm.5].   □

As an immediate consequence, all empirical autocorrelation functions fulfill
$$\sum_\mu \hat\rho(\mu)|\mu| = 0.$$

Simple Random Walks on Landscapes. Before we proceed to landscapes on more general graph let us consider the empirical correlation function along a simple random walk. The geometric relaxation of such a random walk is conveniently described by the probabilities $\varphi_{s\kappa}$, defined in section 2.3.3, that a random walks of length s ends in symmetry class κ. On a vertex transitive graph one obtains [57]
$$\varphi_{s\kappa} = \sum_\mu \frac{|\kappa|\,\rho_\mu(\kappa)}{\langle\rho_\mu;\rho_\mu\rangle}\,\lambda_\mu^s.$$

An analogous formula holds for distance regular graphs, with the distance classes replacing the symmetry classes. For instance, on a sequence space $\mathcal{Q}_\alpha^n$ we obtain explicitly
$$\varphi_{sd} = \sum_{p=0}^n \mathbf{K}_p^{(\alpha,n)}(d)\frac{(\alpha-1)^p\binom{n}{p}}{\alpha^n}\left(1 - \frac{\alpha}{\alpha-1}\frac{p}{n}\right)^s.$$

A simple random walk $\{x_0, x_1, \ldots\}$ induces a time series $\{f(x_0), f(x_1), \ldots\}$ on a landscape f. The autocorrelation function of this time series is defined by
$$\hat{r}(s) \stackrel{\text{def}}{=} \frac{\langle f(x_t)f(x_{t+s})\rangle_t - \langle f\rangle^2}{\langle f^2\rangle - \langle f\rangle^2}.$$

Without losing generality we assume as above that $\langle f\rangle = 0$. Clearly the denominator is just the variance of the landscape, $\sigma^2 = \langle f^2\rangle$. The landscape autocorrelation $\hat\rho(\mu)$ and the random walk autocorrelation are related by
$$\hat{r}(s) = \sum_\kappa \varphi_{s\kappa}\hat\rho(\kappa),$$

see, e.g.,[101]. The relation between the correlation functions $\hat\rho$ and $\hat{r}$ is completely analogous to the relation of ρ and r in section 3.4.3. In particular, we have the following

Theorem 2. *Let Γ be a CGCG with an adjacency matrix with eigenvalues $D\lambda_\mu$ and corresponding left eigenvectors ρ_μ. Then $\hat{r}(s)$ is an autocorrelation function of a simple random walk of a landscape defined on Γ if and only if*
$$\hat{r}(s) = \sum_\mu \Omega_\mu\lambda_\mu^s, \qquad \Omega_0 = 0, \qquad \Omega_\mu \geq 0, \qquad \sum_\mu \Omega_\mu = 1.$$

The coefficients are the same as for the correlation function $\hat\rho$.

Proof. See [49, Thm.6].   □

The eigenvalues of the transition matrix T are explicitly known for sequence spaces:
$$\lambda_d = \frac{\Lambda_d}{D} = 1 - \frac{d}{n}\frac{\alpha}{\alpha-1}.$$

General Vertex Transitive Graphs. Let us denote the set of indices of eigenvectors θ_p belonging to the same eigenvalue $\tilde{\Lambda}_i$ of Δ by $L(i)$. Recall that a vertex transitive graph has the Θ-property if and only if all eigenvalues of its collapsed adjacency matrix are simple

Definition 3. Let Γ be vertex transitive with the Θ-property, and let $\{\theta_y\}$ be a Fourier basis. Then define

$$\omega_i(\mu) = \frac{1}{|\mu|} \frac{\langle \theta_y, R^{(\mu)}\theta_y \rangle}{\langle \theta_y, \theta_y \rangle},$$

where $y \in L(i)$.

Theorem 4. *Let f be a landscape on a vertex transitive graph Γ which has the Θ-property, or a distance regular graph. Then its empirical autocorrelation function is of the form*

$$\hat{\rho}(\mu) = \sum_i \left[\frac{\sum_{q \in L(i)} a_q^2}{\sum_{q \neq 0} a_q^2} \right] \omega_i(\mu)$$

Proof. See [49, Thm.7] for the case of vertex transitive graphs with the Θ-property. The case of distance regular graphs can be dealt with analogously. $\square$

For Cayley graphs of commutative groups we have of course $\omega_i \equiv \rho_i$ for all i. Furthermore, it can be shown that this is also true for a number of examples including the Petersen graph and the Cayley graph $\Gamma(S_5, \mathcal{T})$, see [49]. This, and the very form of the above decomposition, strongly suggests the following

Conjecture 5. *Let Γ be a vertex transitive graph with the Θ-property. Then $\omega_i \equiv \rho_i$.*

We have not been able to find a counter example to this conjecture.

Correlation Length. In order to easily compare different landscapes, or random fields, it is desirable to reduce the information contained in the correlation functions to single number. Most of the empirical correlation functions $\hat{r}(s)$ which have been computed so far are at least approximately decaying exponentials of the form

$$r(s) = \exp(-s/\ell),$$

where the parameter ℓ is the *correlation length*. Numerically, two approaches have been persued in different papers for estimating ℓ, thereby deliberately neglecting potential deviations of $\hat{r}(s)$ from the exponential form. In most cases ℓ is estimated by interpolation from $\hat{r}(\ell_\iota) \stackrel{\text{def}}{=} 1/e$ [23, 20, 24, 22, 32]. Alternatively, one may use linear regression for fitting

$$-\ln \hat{r}(s) \stackrel{\text{def}}{=} \frac{1}{\ell_{lr}} s + c_0,$$

sometimes with the $c_0 = 0$. This approach is used for instance in [101]. Of course, these estimates can differ significantly if there are substantial deviations from an exponential function.

A more elegant definition relies only on the nearest neighbor correlation $r(1)$, using

$$\ell \overset{\text{def}}{=} -\frac{1}{\ln r(1)}.$$

The correlation length — calculated by any of the above numerical procedures — depends linearly on the system size n in almost all model landscapes that have been investigated so far [23, 10, 11, 12, 101], see figure 9. The only notable exception occurs for asymetric TSPs with inversions [10]. This strongly suggests to introduce the *scaled correlation length* $\xi \overset{\text{def}}{=} \ell/n$. It allows to compare entire families of landscapes, as it does not depend on the size n in general.

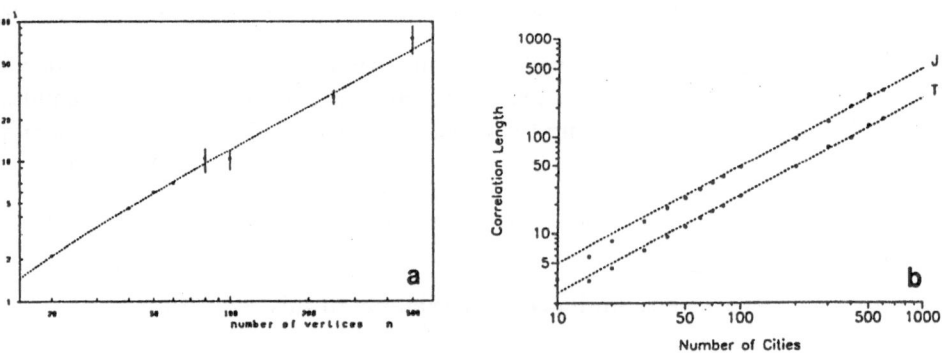

Fig. 9. The correlation lengths of most model landscapes scales linearly with the system size. The examples shown here are: a) Graph Bipartitioning Problem, and b) Travelling Salesman Problem with transpositions (T) and inversions (J), respectively.

Alternative definitions of a scaled correlation function emphasize the geometry of the configuration space graph Γ by comparing ℓ to the diameter of Γ, $\xi' \overset{\text{def}}{=} \ell/\text{diam}\Gamma$, or to the average distance $\bar{d}$ of randomly chosen vertices in Γ, $\xi'' \overset{\text{def}}{=} \ell/\bar{d}$, respectively. We have thus

$$\xi = \frac{n}{\text{diam}\Gamma}\xi' = \frac{n}{\bar{d}}\xi''.$$

The quantities relating the different definitions are, in most cases, numerical constants close to 1. For instance, we have for Hamming graphs

$$\frac{n}{\text{diam}\Gamma} = 1 \quad \text{and} \quad \frac{n}{\bar{d}} = \frac{\alpha}{\alpha - 1},$$

while we both quantities are $1 - \mathcal{O}(\frac{1}{n})$ for the Cayley graphs of the symmetric group with both transpositions and inversions as move sets. The scaled correlation length will play an important role for the estimates of the number of local optima discussed in sect 4.3.2.

4.2 Elementary Landscapes

Definition 6. Let Γ be a graph, Δ the Laplacian of Γ and f a landscape on Γ. We say f is *elementary* if

$$\Delta f + K'(f - \bar{f}) = 0,$$

where $\bar{f} = \langle f \rangle = 1/N \cdot \sum_{x \in \Gamma} f(x)$ is the average value of f, and K' is a constant.

Grover [105] observed that the landscapes of a number of classical combinatorial optimization problems are of this form, see Table 2. In order to keep the notation consistent with Grover's work for regular graphs we introduce $K = K'/D$ where D is as usual the vertex degree. The above relation for f is the discrete analogue of the Helmholtz equation, or reduced wave equation, [106]. An equivalent interpretation is that f is an eigenfunction of the Laplacian operator. It is not surprising that this property is closely related to the properties of the empirical autocorrelation function of the landscapes:

Theorem 7. *Let Γ be a CGCG with vertex degree D. Then the following statements are equivalent:*

(i) f is an elementary landscape with parameter K.
(ii) The empirical autocorrelation function $\hat{\rho}$ of f is a left eigenvector of the collapsed adjacency matrix $\hat{A}$ with eigenvalue $\Lambda = D\lambda = D(1-K) = D-K'$.
(iii) The empirical autocorrelation function $\hat{r}(s)$ of f measured along a simple random walk on Γ is $\hat{r}(s) = \lambda^s$.

Proof. Again we refer to [49]. □

This theorem is a central result in the theory of landscapes. While we have a complete proof only for the special case of Cayley graphs of commutative groups, there is ample evidence that it holds for a much broader class of configuration spaces. In particular it holds true whenever $\rho_i = \omega_i$ for all eigenvalues Λ_i.

A related result linking elementary landscapes to properties of their correlation functions on a larger class of graphs is given below: Let us call $r(1)$ the *empirical nearest neighbor correlation* of the landscape. If Γ is symmetric, that is, if all pairs of neighbors belong to the same orbit, then $\hat{r}(1) = \hat{\rho}(1)$.

Table 2. Elementary Landscapes.

Problem	Move Set	K	$1 - \lambda$
NAES	Hamming	$\frac{4}{n}$	
p-spin	Hamming	$\frac{2p}{n}$	
WP	Hamming	$\frac{4}{n}$	
GC	Hamming	$\frac{2\alpha}{(\alpha-1)n}$	
GBP	Exchange	$\frac{8}{n} - \frac{8}{n^2}$	$\frac{8}{n} - \frac{8}{n^2}$
symmetric TSP	Transposition	$\frac{4}{n}$	$\sim \frac{4}{n}$
	Inversions	$\frac{2}{n-1}$	$\sim \frac{2}{n}$
GMP	Transposition	$\frac{4}{n}$	$\sim \frac{4}{n}$

The configuration space of the GBP is the Johnson graph $\mathcal{J}(n, n/2)$.
The values K for NAES (Non-All-Equal-Satisfyability), WP (Weight Partition), GC (Graph Coloring), GBP (Graph Bipartitioning), and TSP (Traveling Salesman Problem) are taken from [105]. The value of K for the GMP (Graph Matching Problem) is derived in [49]. The values of λ for the GBP and the GMP problem are taken from [12] and [11], respectively; the remaining values are from [20]. The optimization mentionened above are discussed in detail in section 4.4.

Lemma 8. *Let f be an elementary landscape on a symmetric graph Γ. Then $\hat{r}(1) = 1 - K$, with $-DK$ being the eigenvalue of the graph Laplacian corresponding to the landscape.*

Proof. See [49, Lemma 12].   □

4.3 Local Optima

Local Optima and the Eigenstructure of the Laplacian. Local optima are the very feature of a landscape that makes it rugged. An understanding of the distribution of local optima is thus of utmost importance for the understanding of a landscape. The eigenstructure of the graph Laplacian offers a promising formalism for relating the correlation structure of a landscape to the geometry of its local optima. In this section we will briefly describe a few results along these lines.

The geometric structure of an elementary landscape is closely related to the constant K', i.e., the eigenvalue of the graph Laplacian to which f belongs. The solutions of the *Laplace equation* $\Delta f = 0$ form the *harmonic functions* on Γ. It is well known that there are no non-trivial harmonic functions on finite connected graphs (see, e.g., [50]. The harmonic functions correspond to the flat landscapes.

Let f be an arbitrary landscape on Γ. Let

$$V_+ = \{x \in V | f(x) \geq 0\} \quad \text{and} \quad V_- = \{x \in V | f(x) \leq 0\}$$

be the set of all vertices on which f is non-negative or non-positive, respectively. Let Γ_+ and Γ_- be the corresponding induced subgraphs of Γ. The connected components of Γ_+ and Γ_- are called the *nodal domains* of f. Clearly the geometry of the nodal domains is a very important characteristic of the landscape f.

The second largest eigenvalue $\tilde{\Lambda}_1$ of the graph Laplacian, which is negative for all connected graphs, is often called the *algebraic connectivity* of Γ. Theorem 2.5.7 in [107] states that if f is an eigenvector of Δ with eigenvalue $\tilde{\Lambda}_1$, then both Γ_+ and Γ_- are connected, i.e., there are exactly two nodal domains. Following a suggestion by Kauffman we term this type of landscapes "Fujiyama" landscapes as they consist of only a single "mountain" Γ_+. On a Boolean hypercube we have a much stronger result: An elementary landscape with eigenvalue $\tilde{\Lambda}_1$ is of the form

$$f(\sigma) = J_0 + \sum_{k=1}^{n} J_k \sigma_k,$$

and thus there is a unique local maximum $\sigma_{\max}$ and a unique local minimum $\sigma_{\min}$, provided the coefficients J_k are all non-zero. Note also $\sigma_{\min} = -\sigma_{\max}$ in this case. It is unknown if a similar result holds on other graphs as well.

For Riemannian manifolds even more is known. *Courant's Nodal Domain Theorem* states that if all eigenvalues of the Laplacian on a Riemannian manifold are ordered as $0 = \tilde{\Lambda}_1 \geq \tilde{\Lambda}_2 \geq \ldots$, then the number of nodal domains of an eigenfunction θ_k belonging to $\tilde{\Lambda}_k$ is at most k; see, e.g., the book by Chavel [108]. We expect that an analogous theorem holds on graphs. Results along these lines will be presented elsewhere.

Closely related to these observations is theorem 6 of Grover [105]. He showed that the local optima have a characteristic distribution on elementary landscapes: Let $z_{\min}$ and $z_{\max}$ be a local minimum and a local maximum, respectively. Then

$$f(z_{\min}) \leq \bar{f} \leq f(z_{\max}),$$

i.e., if f is an eigenvector of Δ then all local maxima are in Γ_+ and all local minima are in Γ_-. Grover also gives bounds on the length of adaptive walks for elementary landscapes.

The Number of Local Optima. One of the most important characteristics of a landscape is the number $\mathcal{N}$ of local optima. Palmer [109] used an at-least-exponetial increase of $\mathcal{N}$ with the system size n as definition of "rugged" landscape. Local optima are obstacles for heuristic optimization algorithms. It is a very resonable conjecture, therefore, that optimization by a general algorithm should be easier if there are fewer local optima, or probably equivalently, if the nearest neighbor correlation is larger.

It is easy to calculate the number of local optima in two extreme cases: the random energy model, and the completely correlated landscape (in which fitness is added over independent contributions from each element in a sequence). In the completely correlated case, at least in the case of the Boolean hypercube, there is one single optimum. In a random energy model on graph with vertex degree D with N vertices there are on average

$$\mathcal{E}[\mathcal{N}] = \frac{1}{D+1} N$$

local optima. An extension of these results to landscape with a small correlations can be found in [10].

Estimates for the number of local optima are also known for the Nk-model [19, 59, 103], Sherrington-Kirkpatrick model and related short-range 2-spin models [110, 111, 112, 113, 114] for the Travelling Salesman Problem with Transpostion metric [10] and for RNA free energy landscapes [20]. However, there is no routine method for deriving the number of local optima and for relating this number to the corrlation structure of the landscape. In most cases, the number of local increases exponentially with the system size n.

Let us first consider landscapes and random fields on Hamming graphs. Here we have $N = n^\alpha$. In statistical mechanics on considers traditionally the parameters

$$A \stackrel{\text{def}}{=} \lim_{n \to \infty} \frac{1}{n} \ln \mathcal{E}[\mathcal{N}] \qquad \text{and} \qquad A^* \stackrel{\text{def}}{=} \lim_{n \to \infty} \frac{1}{n} \mathcal{E}[\ln \mathcal{N}].$$

It will be convenient to introduce the quantity $P \stackrel{\text{def}}{=} - \lim_{n \to \infty} \frac{1}{n} \ln p_{lo}$, which is related to A via $P = \ln \alpha - A$. Note that we have $\mathcal{N} \approx A^n$ and $p_{lo} \approx P^{-n}$ for large enough systems.

Weinberger [19] gives the following estimate for the probability $p_{lo} = \mathcal{N}/N$ of finding a local optimum in an Nk model for medium and large values of k:

$$p_{lo,Nk} \approx [(\kappa - 1)(k + 1)]^{-\frac{n}{k+1}}.$$

In terms of the scaled correlation length $\xi \stackrel{\text{def}}{=} \ell/n$ this equation becomes

$$P \approx \xi[\ln(\alpha - 1) - \ln \xi].$$

An interesting variation of the Nk model has been introduced recently by Perelson and Macken [115] for modelling the affinity maturation of antibodies by somatic hypermutations. In the *Block Model* a sequence x over an alphabet with α letter is devided into B independent blocks of size B/N^6. The fitness $f(x)$ is defined as the sum of the fitnesses of the individual blocks, which in turn are taken to be i.i.d. random numbers drawn from some distribution with finite variance. The nearest neighbor correlation is given by $r(1) = 1 - 1/B$, implying that the correlation length is approximately $\ell \approx B$ provided $B \gg 1$.

---

[6] In [115] the blocks are not required to have equal size. We impose this restriction here in order to simplify the formalism.

The interesting feature of the block model is that the number of local optima can be easily estimated [115]. One obtains

$$A \approx \frac{1}{n} \ln \mathcal{E}[\mathcal{N}] \approx \ln \alpha - \frac{B}{n} \ln \left((\alpha - 1)\frac{n}{B} \right)$$

and, using $\xi \approx B/n$ for not too few blocks, this translates to

$$A \approx \ln \alpha - \xi[\ln(\alpha - 1) - \ln \xi],$$

and consequently, the probability for hitting a local optimum at random fulfills

$$P \approx \xi[\ln(\alpha - 1) - \ln \xi].$$

This estimate coincides with Weinberger's [19] result for the Nk models.

Using the TAP approach [116] one can estimate the quantities A and A^*. Not unexpectedly, short-rangle spin glasses have more local optima than longe range spin glasses, even the Hamiltonian is of the same form

$$\mathcal{H}(\sigma) = \sum_{(i,j)} J_{ij}\sigma_i\sigma_j$$

where the sum is taken over all pairs (i, j) of neighboring spins [2]. Bray and Moore [111] show that to a first approximation the number of local optima depends on the coordination number z of the spin lattice (i.e., $z = 2$ for a chain of spins, $z = 4$ for a square lattice of Ising spins, and $z = n - 1$ for the SK model.)

$$A(z) \approx A^*(z) \approx 0.1992 + 0.0565/z$$

Exact expressions are available for the 1D case, the linear spin chain:

$$A_{1D} = 4/\pi \approx 0.2415 \quad \text{and} \quad A_{1D}^* = \ln 2/3 \approx 0.2310.$$

These z-expansion is surprisingly accurate: $A(2) \approx 0.2320$. Note that all these 2-spin models have the same empirical correlation function $\hat{\rho}_2$ and thus also the same correlation length $\xi = 1/4$. The numerical values for A from different models, estimated with different approaches are collected in table 3.

Table 3. Estimated values of A for landscapes with $\xi = 1/4$ on Boolean hypercubes.

Model	Method	A	Ref
2-spin SK	TAP	0.1992	[111]
2-spin chain		0.2410	[113]
Nk $k = 3$	direct	0.3466	[19]
Block Model	direct	0.3466	[115]
-	*	0.1308	

The last line is obtained by assuming one local optima in a patch with radius given by the correlation length.

A rough estimate for the number of local optima on landscapes with a Gaussian distribution of values has been proposed in [10]. One concludes immediately from the discussion in sect. 3.7.2 that an assumption on the distribution function is in fact necessary. In a fully correlated landscape, i.e., correlation length $\ell \approx \max d(x, y)$ one expects to find a single optimum. Let $B(r)$ denote the number of vertices in a neighbourhood with radius r around an arbitrary configuration. We expect then, that there are roughly $\mathcal{O}(1)$ local optima in a patch with radius ℓ, i.e., $p_{lo} \approx 1/B(\ell)$. For Hamming graphs this estimate becomes explicitly [29]

$$P = \xi[\ln(\alpha - 1) - \ln \xi] - (1 - \xi)\ln(1 - \xi).$$

It deviates from the above estimates by the term $(1-\xi)\ln(1-\xi)$. See also table 3 for a comparison of numerical values for landscapes with $\xi = 1/4$ on a Boolean hypercube.

The parameters A and P are not useful in cases where the number and/or the probability of local optima is not exponential. The TSP with transposition metric provides such an example. For landscapes on $\Gamma(S_n, \mathcal{T})$ one estimates

$$p_{lo} \approx \frac{1}{\ell!}\left(\frac{e^2}{2\xi^2}\right)^{-\ell} \approx \frac{1}{\Gamma(n\xi)}\left[(e^2/2\xi^2)^\xi\right]^{-n},$$

see [117]. For the TSP with transpositions we have $\xi = 1/4$, i.e.,

$$p_{lo} \approx \mathrm{const}\frac{2.773^{-n}}{(n/4)!}$$

This is in very good agreement with the numerical data [10], see also figure 10.

A rigorous theory linking the correlation structure of Gaussian landscape with the number $\mathcal{N}$ of local optima is still missing. The example of the 2-spin models above shows, furthermore, that there cannot be a simple equation linking the correlation length with the number of local optima. On the other hand, the fairly small numerical range of A for various 2-spin models leaves at least room for rough estimates.

The distribution of local optima cannot be studied directly for large values of n since p_{lo} decreases exponentially in most cases. Some information of local optima, and on the distribution their basins of attraction can be obtained from adaptive and gradient walks, see, e.g., [103]. A statement closely related to the crude estimate for $\mathcal{N}$ is that the length of a gradient walk should be approximately equal to the correlation length, while adaptive walks should be longer by a factor that depends on the structure of the configuration space only. Numerical evidence for the Nk model and RNA landscapes is consistent with this statement.

The statistics of walks and local optima of the random energy model is known in detail [18, 14, 16, 15]. This landscape is qualitatively different from the correlated landscapes discussed above in that the walk lengths are $\mathcal{O}(\log n)$ as opposed to $\mathcal{O}(n)$, and the probability of hitting a local optimum at random is $p_{l.o.} = 1/(1 + D)$, as opposed to exponentially decreasing. Here D is, as usual, the vertex degree of the a configuration, i.e., the number of nearest neighbors.

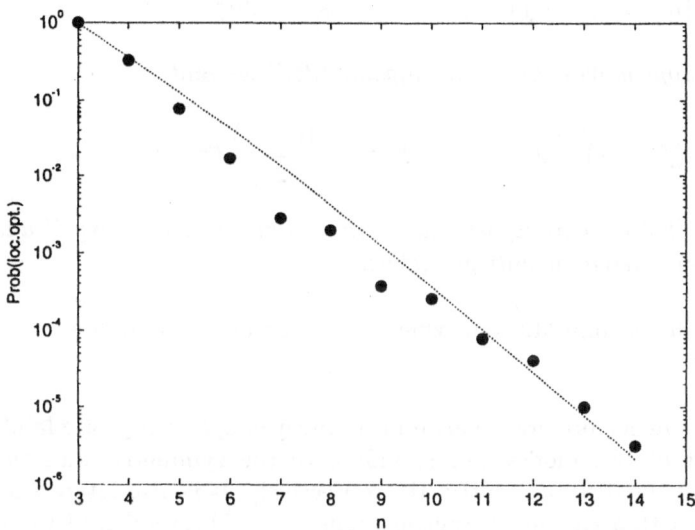

Fig. 10. Number of local optima of a TSP, with neighborhood defined by transpositions, as a function of system size. The dotted line corresponds to one local optimum in a patch with a radius of one correlation length.

4.4 Some Example Landscapes

Traveling Salesman Problems. The Traveling Salesman Problem TSP has already been discussed in the introductory section 2.1. Here we will focus on a few more formal aspects. Recall that the cost of tour π is given by

$$f(\pi) = \sum_j w_{\pi(j)\pi(j-1)},$$

where the indices are interpreted modulo n. An arbitrary matrix W can be uniquely decomposed into its symmetric component $W^\sigma = (W + W^+)/2$ and its antisymmetric component $W^\alpha = (W - W^+)/2$; it will be convenient to define

$$f^\sigma(\pi) \stackrel{\text{def}}{=} \sum_j w^\sigma_{\pi(j)\pi(j-1)} = \frac{f(\pi) + f^*(\pi)}{2},$$

$$f^\alpha(\pi) \stackrel{\text{def}}{=} \sum_j w^\alpha_{\pi(j)\pi(j-1)} = \frac{f(\pi) - f^*(\pi)}{2}.$$

Note that f^σ and f^α can be viewed as cost functions of TSPs with "distance matrices" W^σ and W^α, respectively.

Theorem 9. *Both f^σ and f^α are elementary landscapes on the Cayley graphs of the symmetric group with the transpositions and the inversions as generators, respectively. In particular we have for transpositions*

$$\Delta f^\sigma + 2(n-1)(f^\sigma - \bar{f}) = 0 \qquad \Delta f^\alpha + 2nf^\alpha = 0\,,$$

and for inversions (2opt moves, Lin & Kernighan 1965) we find

$$\Delta f^\sigma + n(f^\sigma - \bar{f}) = 0 \qquad \Delta f^\alpha + \frac{n(n+1)}{2}f^\alpha = 0.$$

The landscape of a TSP with transpositions or inversions is elementary if and only if W is either symmetric or antisymmetric.

Proof. The rather tedious computations which are based on [105] can be found in [49, Thm.13]. □

Asymmetric TSPs hence provide an example of fairly simple composite landscapes. They consist of two modes corresponding to the symmetric and the antisymmetric part of the distance matrix W, respectively. It is also interesting to note in this context that canonical transpositions $(i, i+1)$ do not lead to an elementary landscape.

Consequently, the nearest neighbor correlations of the symmetric and antisymmetric components of a TSP with transpositions are $r(1) = 1 - 4/n$ and $r(1) = 1 - 4/(n-1)$, respectively, i.e., very similar. In fact, numerical estimates [11] had been consistent with $r(1) \sim 1 - 4/n$ for large n in both cases. In the case of inversions we have a symmetric mode with nearest neighbor correlation $r(1) = 1 - 2/(n-1)$ and an antisymmetric contribution with a vanishingly small contribution $r(1) = -2/(n-1) \sim 0$.

It is interesting to correlate these values of $r(1)$ with known facts about the performance of heuristic optimization algorithm, in particular with the simulated annealing. It has been observed by serveral authors that simulated annealing on symmetric TSP is much more effective when reversals instead of transpositions are used as move set, see, e.g., the books by Aarts and Korst [118] or Otten and vanGinneken [119]. Furthermore, Miller and Pekny [38] have observed that reversals are a remarkably bad move set for asymmetric TSPs. These observations are in accordance with the conjecture that landscape with smoother correlation functions have fewer local optima and are thus easier to optimize on, *cp.* [11]. In particular the difference between symmetric and asymmetric TSPs when reversals are used is easily explained in these terms: while for the symmetric TSP the landscape is as smooth as possible, it is completely rugged for the antisymmetric case.

Numerical data by Stadler and Schnabl [11] indicate that the correlation functions $\hat{r}(s)$ of both the symmetric and the antisymmetric components should be of the form $r(1)^s$, we do not, however, have a proof for this conjecture for all n.

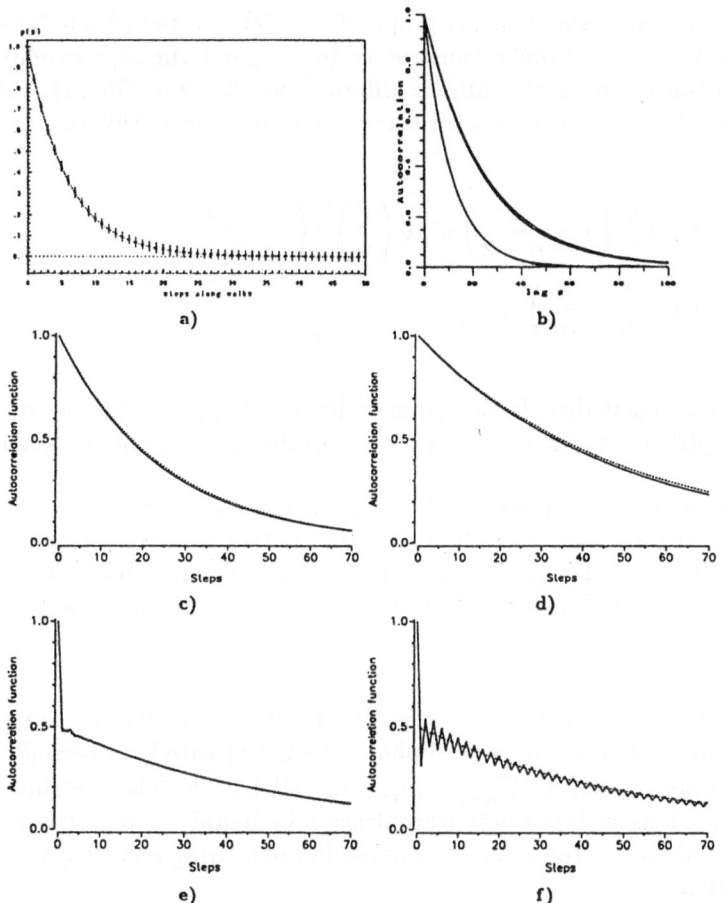

Fig. 11. Examples of autocorrrelation functions.
a) GBP, b) graph matching (upper curves) and low autocorrelated binary string problem, c) TSP with transposition metric, d) symmetric TSP with inversions, e) asymmetric TSP with inversions, f) asymmetric TSP with a restricted set of inversions as move set.

Graph Bipartitioning Problem. Consider a graph G with an even number n of vertices and a symmetric matrix H of edge weights. A configuration is a partition of the vertex set into two subsets A and B of equal size. The cost function is

$$f([A; B]) = \sum_{i \in A} \sum_{j \in B} H_{ij},$$

i.e., the total weight of edges connecting the two subsets. Usually one is interested in minimizing the total edge weight $f([A; B])$ connecting the two subsets A and B, see e.g., [120]. As a close relative of the Sherrington-Kirkpatrick model the *Graph Bipartitioning Problem*, GBP, has received considerable attention.

.The usual move set for this problem is exchanging a vertex in A with a vertex in B. The resulting graph is the Johnson graph $\mathcal{J}(n, n/2)$, see sect. 2.6.3. The random field associated with the GBP is constructed by assigning the edge weights $H_{ij} = H_{ji}$ independently and identically distributed. Stadler and Happel [12] have shown by explicit calculations that the correlation functions of this random field are

$$\rho(d) = 1 - 8\frac{d}{n}\left(1 - \frac{1}{n-2}\right) + 16\left(\frac{d}{n}\right)^2\left(1 - \frac{1}{n-2}\right)$$

$$r(s) = \left(1 - \frac{8}{n} + \frac{8}{n^2}\right)^s,$$

irrespective of the particular distribution from which the H_{ij}'s are drawn. Our results imply that $\rho(d)$ is an eigenvector with eigenvalue $r(1)$ of the Johnson graph $\mathcal{J}(n, n/2)$.

Grover [105] found that each instance of the GBP is an elementary landscape with $K = 8/n - 8/n^2$, see also table 3. Hence we find in fact the expected result $K = 1 - \lambda$. We take this as one further hint that the theorems proved in the previous sections for CGCG are in fact true for a much larger class of configuration spaces.

The Low Autocorrelated Binary String Problem. The LABSP [84, 121] consists of finding binary strings σ over the alphabet $\{-1, +1\}$ with low aperiodic off-peak autocorrelation $R_k(\sigma) = \sum_{i=1}^{N-k} \sigma_i\sigma_{i+k}$ for all lags k. These strings have technical applications such as the synchronization in digital communication systems and the modulation of radar pulses. The quality of a string σ is measured by the fitness function

$$f(\sigma) = \sum_{k=1}^{n-1} R_k(\sigma)^2.$$

In most of the literature on the LABSP the *merit factor* $F(\sigma) = n^2/(2f(\sigma))$ is used. This transformation gives rise to a non-Gaussian distribution of merit factors, see [121] for details.

It can be shown [49, Lemma 16] that the landscape f of the LABSP can be written as

$$f(\sigma) = a_0 + \sum_{k=1}^{\lceil\frac{n}{2}\rceil-1}\sum_{i=1}^{n-2k} 2\varepsilon_{i,i+k}(\sigma) + \sum_{k=1}^{n-1}\sum_{i=1}^{n-1}\sum_{j\neq i-k,i,i+k} \varepsilon_{i,i+k,j,j+k}(\sigma),$$

and thus the empirical autocorrelation functions of the LABS are

$$\hat{\rho}(d) = [1 - \mathcal{O}(1/n)]\rho_4(d) + \mathcal{O}(1/n)\rho_2(d),$$

$$\hat{r}(s) = [1 - \mathcal{O}(1/n)]\left(1 - \frac{8d}{n}\right)^s + \mathcal{O}(1/n)\left(1 - \frac{4d}{n}\right)^s.$$

The landscape of the LABPS is thus not elementary, it consists of a superposition of two modes, namely $p = 2$ and $p = 4$. The smoother $p = 2$ contribution becomes negligible for large n, so that f behaves for long strings almost like an elementary $p = 4$ landscape. This fact explains why the LABPS has been found to be much harder for simulated annealing than, say, the SK spin glass [121]. The empirical autocorrelation function $\hat{r}(s)$ has been computed numerically [11] based on the merit factor F. The numerical estimate for the correlation length

$$\ell \overset{\text{def}}{=} -\frac{1}{\ln r(1)} \approx 0.1230 \cdot n - 0.983$$

is in excellent agreement with the asymptotic value $\ell = n/8 + \mathcal{O}(1)$ obtained above.

Not-All-Equal-Satisfiability. Consider a vector of n binary variable. A literal is a variable or its complement. A clause is a set of three literals that does not contain both a variable and its complement. A clause is said to be satisfied if at least on literal is 0 and at least one literal is 1. An instance of the *Not-All-Equal-Satisfiability Problem* (NAES) is given by a set of c clauses; the cost function is the number of non-satisfied clauses. The move set is defined by flipping the value of a single variable, thus the configuration space is the Boolean Hypercube. It is shown in [105] that NAES has an elementary landscape, see also table 3.

Graph Coloring. An instance of a graph coloring problem GC consists of a graph $G(V, E)$ and a number α of colors. A configuration x is a α-coloring of the vertex set V, i.e., an assignment of one color $x(p)$ to each vertex p of the graph. The cost function is the number edges $(p, q) \in E$ for which both incident vertices p and q have the same color:

$$f(x) = \sum_{(p,q)\in E} \delta_{x(p)x(q)}.$$

A move is the replacement of one color by another one at single vertex. The configuration spaces are thus the General Hamming graphs, i.e., sequence spaces over the alphabet of the α colors. The landscape of a GC is elementary [105], see table 3.

Weight Partition. Given a string of n "spins" $x = (x_i) \in \{-1, +1\}^n$ and corresponding weights w_i, the cost function is given by

$$f(x) = \left(\sum_{i=1}^{n} w_i x_i \right)^2.$$

The move set is given by flipping a single spin, hence the configuration space is again a hypercube. The landscape of a Weight Partition Problem WP is elementary [105], see table 3.

Graph Matching Problem. Given a graph G with n vertices and a symmetric matrix W of edge weights, the task of the Graph Matching Problem GMP is to partition the graph into $n/2$ pairs of vertices such that the sum of the edge weights corresponding to these pairs is optimal. A convenient encoding of the problem is the following. Let $\pi \in S_n$ be a permutation of the vertices. We assume that the vertices are arranged such that $[\pi(2k-1), \pi(2k)]$ form a pair. The cost function is hence

$$f(\pi) = \sum_{k=1}^{n/2} W_{\pi(2k-1),\pi(2k)}.$$

Again, the configuration space is the symmetric group, and hence the set of all transpositions forms a canonical move set. The disadvantage of this encoding is, however, a very large degree of redundancy. In fact, $(n/2)!2^{n/2}$ permutations represent a single matching [122]. As the models discussed above the landscape of the GMP is elementary, see [49, Thm.14] and table 3. Numerical estimates of correlation functions can be found in [11].

4.5 Classification of Landscapes

Fractal Landscapes. Gregory Sorkin [123] proposed the following definition of a fractal fitness landscape which is built on the idea of fractional Brownian motion [124].

Definition 10. A landscape is *fractal* if $\langle \|f(x) - f(y)\|^2 \rangle \propto d^{2h}(x,y)$, where the two configurations x and y with normally distributed fitnesses $f(x)$ and $f(y)$, respectively, are separated by a distance $d(x,y)$.

As a practical matter, the distance between configurations for any finite configuration space is bounded by the diameter of the configuration space diam $\mathcal{C}$, so that Sorkin's definition makes sense only for distances small compared to this upper bound. Sorkin's definition can be reformulated in terms of autocorrelation functions:

$$1 - \hat{\rho}(d) = (1 - r(1))d^{2h}.$$

For simplicity we will assume in this section that the autocorrelation functions are given in terms of distances, otherwise we may work with the corresponding random walk version. The details of this step are explained in the following section 4.7. For distances d small compared to the correlation length ℓ, almost all of the landscapes listed in table 4, as well as most of the RNA landscapes described in section 5, are of the form

$$\rho(d) = 1 - \frac{d}{\ell} + \dots,$$

and therefore approximately satisfy Sorkin's definition with $h = 1/2$.

Table 4. Combinatorial optimization problems and their autocorrelation functions.

Name	Metric	$\rho(d)$	$r(s)$	ℓ	diamΓ	Ref.
REM	any	$\delta_{0,d}$	see 3.6.2	0	n	-
s.TSP	Tr.	?	$\sim e^{-4s/n}$	$n/4$	$n-1$	[10]
	2opt	?	$\sim e^{-2s/n}$	$n/2$	$n/2 \ldots n-1$	[10]
	c.Tr.	?	$\sim e^{-2s/n}$	$n/2$	$\frac{n(n-1)}{2}$	[21]
a.TSP	Tr.	?	$\sim e^{-4s/n}$	$n/4$	$n-1$	[10]
	2opt	?	$\sim \frac{1}{2}(\delta_{0,s} + e^{-2s/n})$	-	-	[10]
	c.Tr.	?	$\sim e^{-3s/n}$	$n/3$	$\frac{n(n-1)}{2}$	[21]
GM	Tr.	?	$\sim e^{-4s/n}$	$n/4$	$n-1$	[11]
GBP	Ex.	$1 - \frac{n-1}{n-2}[8\frac{d}{n} - 16(\frac{d}{n})^2]$	$(1 - \frac{8}{n} + \frac{8}{n^2})^s$	$(n-3)/8$	$n/2$	[12]
LAS	Ham.	?	$\sim e^{-10s/n}$	$n/10$	n	[11]
rnd.Nk	Ham.	$(1 - \frac{d}{n})(1 - \frac{k}{n-1})^d$	$\sim (1 - \frac{k+1}{n})^s$	$n/(k+1)$	n	[19, 20]
p.r.Nk	Ham.	$(1 - \frac{k+1}{n})^d$	$\sim (1 - \frac{k+1}{n})^s$	$n/(k+1)$	n	
adj.Nk	Ham.	see 3.6.3	$\sim (1 - \frac{k+1}{n})^s$	$n/(k+1)$	n	[19, 20]
p-Spin	Ham.	$1 - \frac{2}{\binom{n}{p}} \sum' \binom{d}{j}\binom{n-d}{p-j}$	$\sim e^{-2ps/n}$	$n/(2p)$	n	[21]
SK	Ham.	$1 - \frac{n}{n-1}[4\frac{d}{n} - 4(\frac{d}{n})^2]$	$(1 - \frac{4}{n})^s$	$n/4$	n	[21]

$\sum'$ for the p-spin model denotes the sum over all odd j subject to the restriction $j > \min(d,p)$. REM is Derrida's random energy model; s.TSP and a.TSP denote symmetric and asymmetric travelling salesman problems [34], GM is the graph matching problem [125]. The corresponding metrics are transpositions (Tr.), 2opt moves and canonical transpositions (c.Tr.). GBP is the graph bipartitioning problem [120], its metric (Exc) is derived from exchanging a pair of objects. LAS stands for the low autocorrelated string problem [84, 121]. The Sherrington-Kirkpatrick spinglass [97] is the special case $p=2$ of the p-Spin model [96, 98, 99] introduced in [8] as a model for a rugged landscape in evolutionary optimization. The abbreviations rnd.Nk, p.r.Nk, and adj.Nk refer to random neighbour, purely random and adjacent neighbour Nk-model, resp. Here the canonical metric is the Hamming metric.

Classification by the "Form" of the Correlation Functions. Since strictly speaking all landscapes are finite objects a classification beyond the distinction elementary/non-elementary seems to be hard to achieve. Instead of considering a single landscape, we assume therefore that we are given a family $\{f_n\}$ of landscapes (or random fields) on a graphs Γ_n. The index $n = 1, 2, \ldots$ is a "natural" measure of the size of the configuration space. For sequence spaces, for instance, n is simply the chain length. Now consider the correlation functions $\rho_{[n]}(d)$ and $r_{[n]}(s)$ of the landscape f_n. We define the scaled autocorrelation functions $\zeta(h)$ and $z(h)$ for all rational numbers h by

$$\zeta(h) = \lim_{n \to \infty} \rho_{[n]}(nh) \qquad z(h) = \lim_{n \to \infty} r_{[n]}(nh),$$

respectively. Note that ζ and z do not necessarily exist! An example is the family of p-spin models with $p = n$ for which we find $\rho_{[n]}(n/2) = \pm 1$ depending on whether $n \equiv 0 \bmod 4$ or $n \equiv 2 \bmod 4$.

Based on the "shape" of $\zeta(h)$ or $z(h)$, which the same, Weinberger and Stadler [21] proposed a classification of landscapes. This work was motivated by the analogy to the classification of stationary continuous-time stochastic processes in terms of their autocorrelation functions.

I $\zeta(h)$ is discontinuous for $h = 0$. These landscapes are *extremely rugged*. (A continuous time stochastic process with an autocorrelation function of this type is not even continuous.) Two subclasses can be distinguished:

Ia $\zeta(h) = \delta(h)$ (or more generally consists of a finite sum of delta functions). This subclass may be divided further depending on the behaviour of the unscaled correlation function:

* The unscaled correlation length ℓ diverges when diam $\mathcal{C} \to \infty$. An example of this subclass is the TSP with canonical transpositions defining the neighborhood. The correlation length is $\mathcal{O}(n)$, whereas the diameter of the landscape is $\mathcal{O}(n^2)$ where n is the number of cities; see table 4.

* The unscaled correlation length ℓ is uniformly bounded from above. Examples of this subclass are the random energy model and (probably) certain RNA landscapes [23].

Ib $\zeta(h)$ does not vanish in a neighbourhood of $h = 0$ but the autocorrelation function is bounded from above by some constant $\gamma < 1$ for $h \neq 0$ in a neighborhood of $h = 0$, i.e., there is a finite size jump at $h = 0$, but the landscape is not uncorrelated. An example is provided by the asymmetric TSP with 2-opt moves, see figure 11.

II $\zeta(h) = 1 - \alpha h + \mathcal{O}(h^2)$, with a constant α of order 1. Examples are almost all entries in table 4. Class **II** landscapes are "trivial" random fractals with $h = \frac{1}{2}$. Continuous-time stochastic processes with an autocorrelation function of this type are continous but not differentiable; hence these landscapes are "rugged".

III $\zeta(h) = 1 - \beta h^2 + o(h^2)$, with some constant β of order 1. In contrast to all previous cases the autocorrelation function is differentiable at $h = 0$.

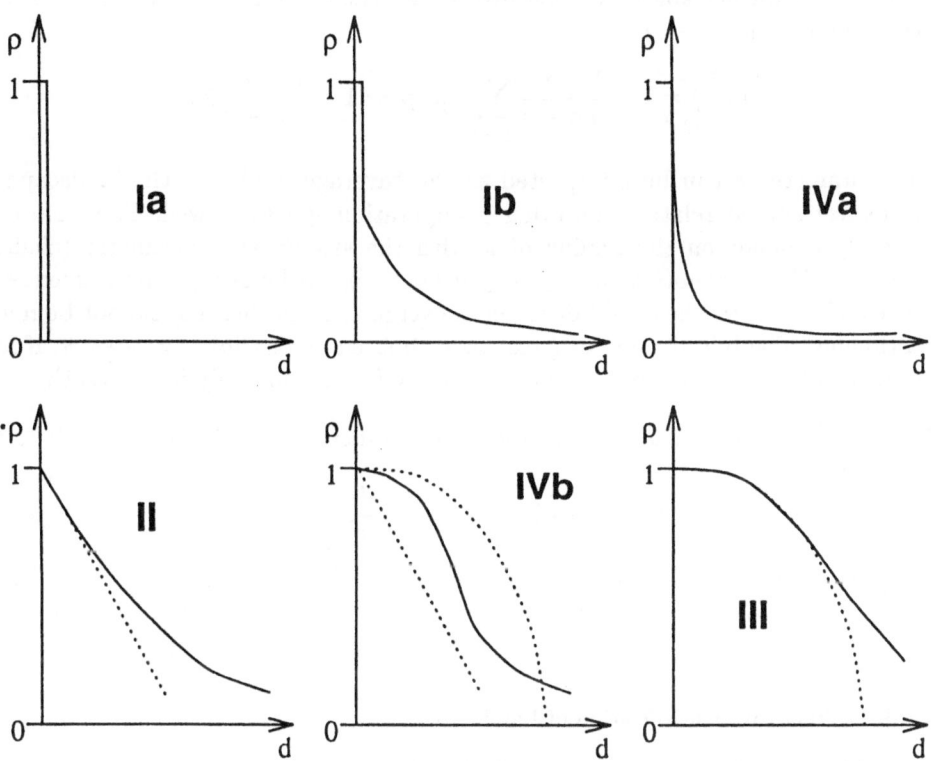

Fig. 12. Classification of landscapes (and random fields) according to the "shape" of their correlation functions. For details see text.

Correlation functions of this type belong to differentiable stochastic processes in the continuous-time case. These landscapes form the second class of trivial fractals, corresponding formally to Sorkin's definition of fractals with $h = 1$. These are the smoothest possible landscapes, since any function $f(x) = 1 - \mathcal{O}(x^{2+q})$ with $q > 0$ cannot be autocorrelation function of a stochastic process.

IV $\zeta(h) = 1 - c_p h^p + o(h^p)$, with $p \neq 1, 2$, correspond to non-trivial random fractals. Such landscapes have been constructed by Sorkin on fractal config-uration spaces, but have not (yet?) been found in "real life" examples.

Classification Based on the Characterization of Correlation Function.
The characterization of autocorrelation functions has direct implications on the *classification* of autocorrelation function. We will now relate the classes defined

in the previous subsection with the spectral decomposition of the autocorrelation function described in section 4.1.

On the sequence spaces we can investigate the behavior of $\zeta(h)$ for small h explicitly. Consider

$$\zeta(\frac{1}{n}) = 1 - \frac{1}{n}\frac{\alpha}{\alpha - 1}\sum_{p=0}\Omega_p \cdot p =: 1 - \frac{1}{n}\frac{\alpha}{\alpha - 1}\bar{\pi}.$$

The parameter $\bar{\pi}$ can be interpreted as the "average mode" of the landscape. We expect similar relations on other configuration spaces as well. The form of ζ clearly depends on the scaling of $\bar{\pi}$ with the system size parameter (chain length) n. First we note that $1 \leq \bar{\pi} \leq n$ holds for all landscapes on sequences spaces. If $\bar{\pi} = \mathcal{O}(n)$ then $\zeta(\frac{1}{n})$ does not converge to 1 and hence ζ cannot be not continuous. The initial slope of ζ can be estimated as $\partial\zeta/\partial h|_{0+} = -\frac{\alpha}{\alpha-1}\bar{\pi}$, if it exists at all. Hence we have an asymptotically finite slope only if $\bar{\pi} = \mathcal{O}(1)$.

Theorem 11. *The smoothest landscape on a sequence space has autocorrelation function*

$$\rho_1(d) = 1 - \frac{\alpha}{\alpha - 1}\frac{d}{n}.$$

Proof. See [49].   □

Table 5. Tentative Classification of Landscapes.

	Class I	rugged "Fractals"	Class II	smooth "Fractals"	Class III
ζ	not continuous	$\zeta = 1 - c|h|^\gamma$ $0 < \gamma < 1$	$\zeta = 1 - c|h|$	$\zeta = 1 - c|h|^\gamma$ $1 < \gamma < 2$	$\zeta = 1 - c|h|^2$
$\bar{\pi}$	$\bar{\pi} = \mathcal{O}(n)$	$\bar{\pi} \sim o(n), \bar{\pi} \to \infty$	$\bar{\pi} = \mathcal{O}(1)$	do not exist on $\mathcal{Q}_\alpha^n$	
	Class A	Class B	Class C		

This reasoning suggests that there should be three different classes of landscapes:

A $\bar{\pi} = \mathcal{O}(n)$ implies $\lim_{n\to\infty}\zeta(\frac{1}{n}) < 1$ and hence the asymptotic autocorrelation function is discontinuous at the origin.

B $\bar{\pi} \to \infty$ but $\bar{\pi} = o(n)$ allows ζ to be continuous with infinite initial slope.

C $\bar{\pi} \sim \mathcal{O}(1)$ implies an autocorrelation function which decays continuously with finite slope.

We do *not* claim that there is one-to-one correspondence between the two classification schemes, which are compared in table 5. *One of the reasons for this is that we have no general conditions for the existence of ζ.*

4.6 Landscapes on Irregular Graphs

Correlation Functions. Let us use the notation $D_h(x) \stackrel{\text{def}}{=} |N_h(x)|$ for the distance degrees, i.e., for the number of vertices in distance h from a vertex x. Suppose Γ be a graph with N vertices and M_h unordered pairs of vertices with mutual distance h. Of course M_1 is the number of edges. Interpreting D_h as a landscape we obtain immediately the balance equation

$$\sum_x D_h(x) = \langle D_h \rangle N = 2M_h.$$

On non-regular graphs we have essentially two ways of defining a correlation function.

Definition 12. Let $f : V\Gamma \to \mathbb{R}$ be a non-constant landscape on a connected graph Γ with $N > 1$ vertices. Then we define

$$\hat{\rho}(d) \stackrel{\text{def}}{=} \frac{\langle f(x)f(y) \rangle_{d(x,y)=d} - \langle f \rangle^2}{\langle f^2 \rangle - \langle f \rangle^2}$$

$$\hat{\rho}_*(d) \stackrel{\text{def}}{=} 1 - \frac{\langle (f(x) - f(y))^2 \rangle_{d(x,y)=d}}{\langle (f(x) - f(y))^2 \rangle_{(x,y)\in V \times V}}$$

Both definitions of correlations have their advantages in certain contexts. $\hat{\rho}$ is the usual definition which have already encountered for distance regular graphs. $\hat{\rho}_*$ on the other hand has the same flavour as the usual correlation coefficient. Since it depends only on squared differences it will form the starting point for a generalization of correlation measures to non-numeric functions in sect. 5.2., see also [20, 24]. In fact, two definitions coincide provided the configuration space is sufficiently symmetric:

Theorem 13. *Let Γ be distance degree regular. Then $\hat{\rho} \equiv \hat{\rho}_*$*

Proof. See [20]. □

Definition 14. Let f and g be two landscapes on Γ. We define the *covariance* of f and g by

$$\text{cov}[f, g] \stackrel{\text{def}}{=} \langle fg \rangle - \langle f \rangle \langle g \rangle.$$

Note that this conforms the usual definition of a covariance.

For each non-constant landscape on a connected graphs we define the function

$$\Psi(h) \stackrel{\text{def}}{=} \frac{\text{cov}[D_h, \varphi_f]}{\langle D_h \rangle \langle \varphi_f \rangle} \qquad \text{where} \qquad \varphi_f(x) \stackrel{\text{def}}{=} f^2(x) - \langle f \rangle^2.$$

With this definition we can prove an interested generalization of the above relation between $\hat{\rho}$ and $\hat{\rho}_*$.

Theorem 15. $\rho_*(h) - \rho(h) = -\Psi(h)$ *for all connected graphs and all non-constant landscapes.*

Proof. The proof proceeds by a straight forward (although rather tedious) direct evaluation of both sides of the equation. We omit the details here. □

Note that $\Psi(0) = 0$ for all graphs, and $\Psi(1) = 0$ holds for all landscapes on regular graphs. We suspect that landscapes for which Ψ vanishes for all distances h will play a special role on irregular graphs. We propose to call such landscapes *uniform.* Note also that by replacing the landscape averages by expectation values one obtains an analogous formula relating the correlation function $\rho(d)$ on a random field and it "squared difference" counterpart $\rho_*(d)$.

An Upper Bound on the Nearest Neighbor Correlation.

Theorem 16. *Let $\tilde{\Lambda}_1$ be the second eigenvector of the Laplacian Δ of Γ, let f be an arbitrary landscape on Γ with correlation function $\hat{\rho}_*$. Then*

$$\hat{\rho}_*(1) \leq 1 - \frac{1}{\langle D \rangle}\tilde{\Lambda}_1.$$

Proof. We start with an observation by Biggs [46, p.58], namely

$$\tilde{\Lambda}_1 \leq \sum_{\{x,y\}\in E} (f(x) - f(y))^2 \Big/ \sum_{x\in V\Gamma} f^2(x)$$

This can be rewritten as $\tilde{\Lambda}_1 \leq \frac{M}{N\sigma^2}\langle (f(x) - f(y))^2\rangle_{d(x,y)=1}$. Multiplying by $2N/M = 1/\langle D \rangle$ yields

$$\lambda_1 = \frac{\tilde{\Lambda}_1}{\langle D \rangle} = \frac{\langle (f(x) - f(y))^2\rangle_{d(x,y)=1}}{2\sigma^2} = 1 - \hat{\rho}_*(1).$$

A simple rearrangement completes the proof. □

Corollary 17. *If Γ is regular with vertex degree D, then we have*

$$\hat{\rho}(1) = \hat{\rho}_*(1) \leq \frac{\Lambda_1}{D} = \lambda_1,$$

where Λ_1 is the second largest eigenvector of the adjacency matrix A and λ_1 is the second largest eigenvector of the transition matrix of a simple random walk on Γ [57, 95].

Proof. This follows immediately from the definition $\Delta = D - A$ and the fact that D is a multiple of the identity matrix in case of regular graphs. □

This theorem effectively gives an upper bound on the nearest neighbor correlation of arbitrary landscapes on arbitrary configuration spaces. It provides also an alternative proof for the theorem in section 4.5.3.

4.7 Empirical Anisotropy

It is crucial for an understanding of the dynamics of a (heuristic) optimization procedure to know whether the landscape looks like a typical instance of an isotropic random field or whether there are significant anisotropies. The great importance of the ridge-like anisotropies for the dynamics of evolutionary adaptation, for instance, is discussed in [9].

The random field definition of isotropy becomes useless if we are given only a single instance. It is clear that we can only base our approach on a comparison of samples taken from different regions of the configuration space C. One has to be careful when choosing these samples, however, since we have to expect uncontrolled effects on the sample statistics if we average over set of configurations with unequal geometries. For example, the mean fitness along a "straight line" in configuration space will in general be quite different from the mean fitness of a configuration and its neighbors. Recently, a collection B of test sets with the following four properties has been proposed as a suitable sampling strategy [31]

(i) B is a partition of the configuration space C.
(ii) $A \in B$ is a connected subgraph of C.
(iii) Any two subgraphs $R, S \in B$ are isomorphic, i.e., they have the same geometry.
(iv) $1 \ll |A| \ll |C|$, for all $A \in B$.

The first three requirements make sure that we take fair samples. Note that these conditions are quite restrictive. The last condition ensures that we have enough samples. The basic definition in [31] is the following:

Definition 18. A value landscape is *empirically isotropic* if for any family B of test sets holds
$$\left\langle \left(f(x) - f(y)\right)^2 \right\rangle_\mu^A = \left\langle \left(f(x) - f(y)\right)^2 \right\rangle_\mu,$$
i.e., the average of the squared fitness difference of pairs of configurations belonging to a given symmetry class μ will be the same whether it is measured in single test set or over the entire configuration space. By replacing the symmetry classes by distance classes one obtains a completely analogous definition for *empirically homogeneous* landscapes.

The average correlation of pairs of configurations in a test set, $\bar{\rho}_B$, will not vanish in general. This reflects that the test sets A are much smaller than the entire configuration space C. It can be calculated from the empirical autocorrelation functions discussed earlier in this chapter:

$$\bar{\rho}_B = \sum_\mu p_A(\mu)\hat{\rho}(\mu) \qquad \text{or} \qquad \bar{\rho}_B = \sum_d p_A(d)\hat{\rho}_h(d)$$

where $p_A(\mu)$ and $p_A(d)$ are the probability that a randomly chosen pair of configurations from the test set A belongs to the symmetry class μ and to the distance class d, respectively.

Now consider the variance $\mathrm{var}_{\mathcal{B}}[\langle f \rangle_A]$, measured over all test sets $A \in \mathcal{B}$, of the average fitness values $\langle f \rangle_A$ within the test sets A. The main technical result in [31] is the following

Theorem 19. *Let f be an empirically isotropic (empirically homogeneous) landscape, let $\mathcal{B}$ be a collection of test sets fulfilling (i) through (iv), and let $\bar{\rho}_{\mathcal{B}}$ be defined as above via the symmetry classes (distance classes) of the configuration space. Then*

$$\mathrm{var}_{\mathcal{B}}[\langle f \rangle_A] = \sigma^2(1 - \bar{\rho}_{\mathcal{B}}),$$

where σ^2 is as usual the variance of all fitness values.

Proof. See [31]. □

It seems natural to measure the anisotropy of a landscape by the extent to which this relation is violated. We therefore call the dimensionless parameter

$$\alpha_{\mathcal{B}} = \frac{\mathrm{var}_{\mathcal{B}}[\langle f \rangle_A]}{\sigma^2} - \bar{\rho}_{\mathcal{B}}$$

the *coefficient of anisotropy* with respect to the partition $\mathcal{B}$. From a theoretical point of view it is tempting to define the "true anisotropy" of a landscape as the maximum value of $\alpha_{\mathcal{B}}$ from all partitions $\mathcal{B}$. From a practical point of view, however, one would need prohibitively large computer resources to actually compute $\max_{\mathcal{B}} \alpha_{\mathcal{B}}$. A simple application is shown as figure 13.

The main advantage of this approach is that it is as close as possible to the definitions for random fields. Moreover, all quantities have the flavor of second moments and can be computed quite efficiently. Its major disadvantages are the dependence of the numerical values on the choice of the collection of test sets $\mathcal{B}$, and the fact that it gives only global information, thereby neglecting the overwhelming importance of the geometry of the anisotropies.

5 Biological Landscape: RNA

5.1 RNA Secondary Structures

Structure Formation. RNA molecules spontaneously form three-dimensional structures by folding the sequences in aqueous solutions, which contain appropriate concentrations of structure stabilizing divalent cations like Mg^{2+}, and have appropriate ionic strength, pH and temperature. The major driving force for structure formation is Watson-Crick base pairing ($\mathbf{G}{\equiv}\mathbf{C}$, and $\mathbf{A}{=}\mathbf{U}$) mediated by partial intramolecular complementarity of sequences, as well as $\mathbf{G}{-}\mathbf{U}$ base pair formation. Other intermolecular forces and the interaction with the aqueous solvent shape the spatial structure of an RNA molecule. Small and medium size RNA molecules (with chain lengths $n < 200$) form equilibrium structures which are independent of the mechanism of folding and thus are completely determined by the sequence and the environmental conditions. Structure formation of large

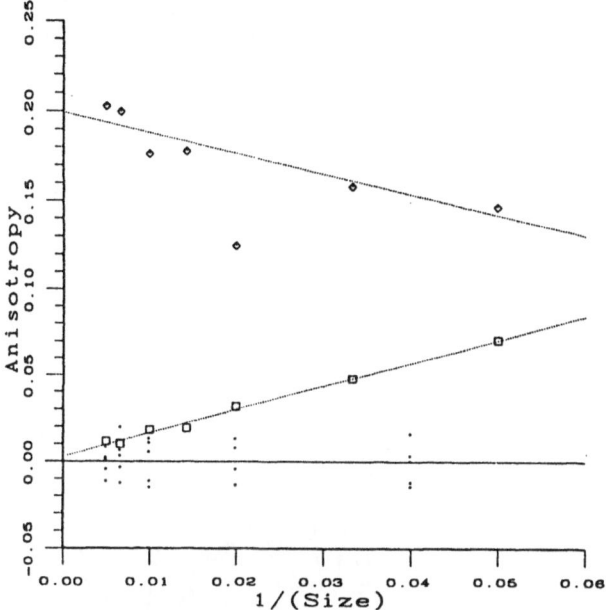

Fig. 13. Empirical coefficient of anisotropy as a function of chain length for three model landscapes.

($\cdot$) Sherrington-Kirkpatrik spin glass with $2n$ spins. A reasonable collection of test sets is obtained by fixing one halve of the spins within each given test set, which is thus a hypercube of dimension n. The corresponding random field is is isotropic (see sect 3.4.1). The numerical data indicate that the instances of the SK model are indeed empirically isotropic.

In contrast the two examples of RNA free energy landscapes (see section 5 for details) exhibit anisotropies. The RNA free energy landscape of the natural four-letter alphabet **GCAU** and the landscape of the artificial **GCXK** alphabet differ in the logic and strenght of the possible base pairs. In both cases the test sets are constructed as follow: A_σ is the set of all sequences that have **G** or **C** at each position i with $\sigma_i = 1$ and **A** or **U** (or **X** or **K**, respectively) on the $\sigma_i = 0$ positions. The binary string σ thus characterizes the "slice" A_σ (which is a Boolean Hypercube of dimension n). The data show that the anisotropies vanish asymptotically for the artificial **GCXK** alphabet ($\square$) which has two equivalent types of base pairs (**GC** and **XK**). In constrast, we find a large amount of anisotropy for the natural alphabet, for which the base pair **AU** is much weaker than **GC** and there is the possibility of forming **GU** pair ($\diamond$).

RNA molecules appears to be controlled at least in part by the kinetics of folding process.

Formation of spatial structures of RNA molecules is commonly partitioned into two steps (Figure 14). The first step comprises conventional base pair formation (**AU**, **GC**, **GU**) and yields the so-called secondary structure of the RNA molecule, see below. In the second step the base pairing pattern is converted into the 3D structure.

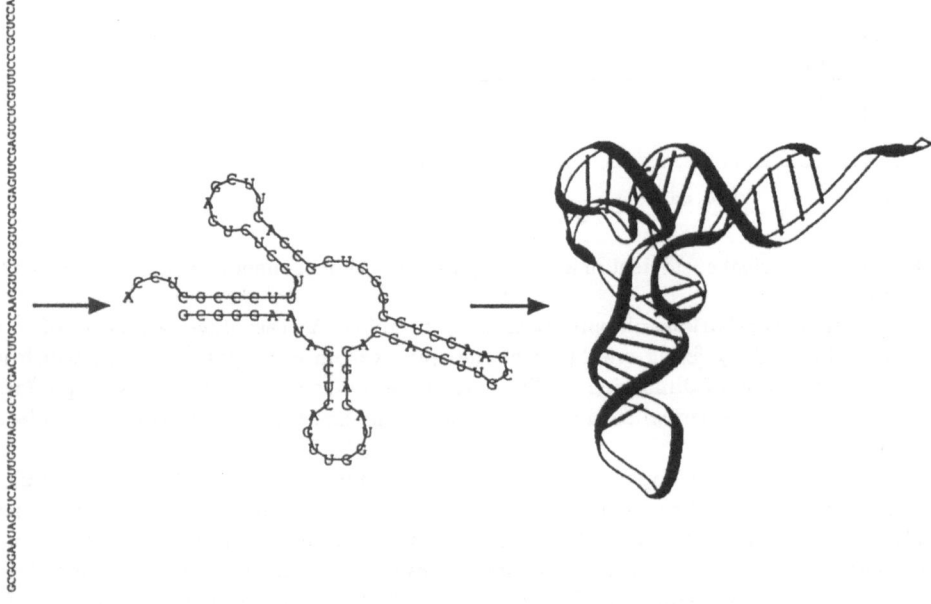

Fig. 14. Folding of an RNA sequence into its spatial structure. The process is partitioned into two phases: in the first phase only the Watson-Crick-type base pairs are formed (which constitute the major fraction of the free energy), and in the second phase the actual spatial structure is built by folding the planar graph into a three-dimensional object. The example shown here is phenylalanyl-transfer-RNA (t-RNA [phe]) whose spatial structure is known from X-ray crystallography.

Structure Prediction. Secondary structures are defined as listings of the Watson-Crick-type base pairs in the actual structure which fulfill the "no-knot condition" and may therefore be represented as planar graphs. Two features of RNA secondary structures will be important later on. They are discrete by definition (two bases either form pair or they don't), and they are composed of largely independent structural elements:

- *stacks*, i.e., double-helical regions,
- *loops*, i.e., unpaired regions enclosed by stacks (the degree of a loop is the number of stacks attached to it; consequently, a hairpin loop has degree 1, bulges and interior loops have degree 2, and all loops with degrees larger than 2 are multi-loops), and
- *external elements* are strains of unpaired bases that are not part of a loop (they are either joints, which connect components, or free ends).

There are several reasons for considering the secondary structure as a crude first approximation to the spatial structure of the RNA.

- Conventional base pairing and base pair stacking cover the major part of the free energy of folding.
- Secondary structures are used successfully in the interpretation of RNA function and reactivity.
- Secondary structures are conserved in evolutionary phylogeny.

The statistical investigation of RNA based landscapes requires the knowledge of several hundred thousand values derived from RNA structures. These data are not available at present, neither through experimental measurement nor through computation of the 3D-structures (which are highly time consuming and still unreliable). Restriction to secondary structures as a crude approximation to real structures, however, renders computation possible.

A variety of computer programs predicting RNA secondary structures have been published. A very brief overview is given in Table 6. Two public domain packages for RNA folding are currently available by anonymous ftp: Zuker's mfold [134] and the **Vienna RNA Package** [135], the latter providing a variety of different algorithms for structure prediction and structure comparison [25].

5.2 RNA Free Energy Landscapes

The most obvious biophysical quantity to investigate is simple the stability of a (secondary) structure, that is, its free energy of structure formation. The RNA free energy landscapes may serve as a prototype for biophysically interesting landscapes despite that fact that there is no evolutionary process that would optimize the free energy of structure formation. It is has been found recently, for instance, that the landscapes of activation energies for the melting of secondary structures closely resemble the corresponding minimum free energy landscapes [32]. Free energy landscapes have been computed with a variety of algorithms (see table 6) for a number of different alphabets (see table 5.2).

Table 6. Folding Algorithms.

Algorithm	ψ	Abbr.	Remark	Reference
deterministic				
Minimum Free Energy	-	MFE	fast	[126, 127]
Kinetic Folding	+	KIN	fast	[128]
5'-3' Folding	+	5-3	fast	[129]
Partition Function	-	PF	ensemble	[130]
Maximum Matching	-	MM	unrealistic	[131]
stochastic				
Simulated Annealing	+	SA	very slow	[132, 133]

ψ Pseudo-knots can be included. The major problem with the prediction of pseudo-knots is, however, the lack of sufficient experimental energy parameters.

Correlation lengths of free energies are essentially linear functions of the chain length n. Numerical data of scaled correlation lengths are compiled in table 5.2. The base pairing alphabet has remarkably strong influence on the correlation length: the correlation length for **AUGC** landscapes are much longer than for pure **GC** or pure **AU**-sequences. Detailed information on the RNA free energy landscapes can be found in [23, 20, 24].

As an application of the Fourier-theory of landscapes we briefly discuss here the decomposition of RNA free energy landscapes into their modes. The correlation data are taken from [20]. The data are fitted reasonably well by by a two-mode model of the form $\hat{\rho} = a\rho_k + (1 - a)\rho_l$. The coefficient $0 \leq a \leq 1$ and the modes k and l cannot be computed directly due to the fairly large amount of noise in the empirical correlation data. We therefore had to resort to a least square fitting procedure that takes the non-negativity of the coefficient(s) explicitly into account and uses only a small number of non-zero coefficients. We do not expect the RNA free energy landscapes to be *exactly* superpositions of only two modes, we rather expect a broad peak of the spectrum concentrated at about $\ell = 5 \ldots 8$. Using more than two modes increases the quality of the fits only slightly. Furthermore the fits with more than two coefficients do not compare well when independent data sets are used. Numerical data are given in table 5.2. We conclude hence that our numerical estimates of the correlation functions $\hat{\rho}(\mu)$ are not accurate enough (approximate noise level $\pm 5\%$) for obtaining a more detailed decomposition of the RNA free energy landscapes.

We find a pronounced difference between the natural **GCAU** alphabet and an artificial alphabet **GCXK** with two base pairs, **GC** and **XK**, of equal strength. While the latter is represented very well by the $\ell = 6$ mode, we obtain a superposition of the modes $\ell = 1$ and $\ell = 6$ in case of the natural alphabet. This does not come as a surprise, however. There is an essentially linear

Table 7. Scaled correlation length ℓ^* for RNA free energy landscapes.

	ΔG				ΔG^{ne}		Model†	
Alphabet	MFE	MFE[s]$^+$	PF	KIN	Melt.	Rec.	D	A
GC	0.0857	0.0768	0.127	0.0448	0.099	0.030	$\leq 3.5^*$	≤ 3 *
AU	0.0600							
GU	0.0514							
AUI	0.0532							
AGC	0.0982							
AUC	0.0856							
AUG	0.1191							
UGC	0.3850							
GCAU	0.2627	0.2976	0.243	0.1734	0.268	0.055	0.2	$\leq 15^*$
GCAU$^\%$	0.2044							
GCXK	0.1070				0.125	0.070		
ABCDEF	0.1182							

All data refer to $T=37^\circ$C.
$^+$ [s] refers to Salser's old parameter set [136], all other data have been obtained with a recently updated parameter set [137].
* A linear dependence on the chain length could not be verified. The bounds correspond to chainlength $n \leq 120$.
† These landscapes have been used to model replication and degradation in the computer simulations [6, 7].
$^\%$ GU pairs forbidden.

increase of the free energy with increasing **GC** content which accounts for the $\ell = 1$ mode. (Recall that $\ell = 1$ corresponds to the additive fitness model, or 'Fujijama' landscape). The order of the dominating modes allows one to draw conclusions on the approximate size of the building blocks underlying the landscape: in case of RNA free energies the data indicate that on average a nucleotide interacts strongly with about 5 other position, leading to a building block size of 6. This is a perfectly reasonable number given the energy model [127] for RNA secondary structure formation: a major contribution of the energy comes from

Table 8. Decomposition of RNA Free Energy Landscapes.

n	a_6			a_1		a_7
	GCAU	GCXK	GC	GCAU	GCXK	GC
20	0.68	0.86		0.32	0.14	
30	0.65	0.88		0.35	0.12	
50	0.57	0.93	0.73	0.43	0.07	0.27
70	0.57	0.93		0.43	0.07	
100	0.53	0.96		0.47	0.04	

stacking of base pairs, hence we expect at least the pairing partners and their neighbors to be important.

5.3 Combinatory Maps

The Metric Space of RNA Secondary Structures. The dichotomy between genotypes and phenotypes is basic to biological evolution. Molecular biology has shown that the genetic information for the unfolding of organisms, which represent the phenotypes, is contained in the DNA, which is the genotype. RNA molecules in cell-free evolution experiments are both genotypes and phenotypes. Following Sol Spiegelman we consider the spatial structure of the RNA molecule as its phenotype. Indeed, this structure is evaluated by the selection process. The structure of the RNA molecule acts thus as mediator between the sequence and the scalar property that is considered in the (fitness) landscape. The relation between sequences and structures thus seems to be basic to all landscapes of biopolymers.

The *shape space* $\mathcal{Y}$ is defined as the set of all possible RNA secondary structures. The notion of a shape space was used previously in theoretical immunology for the set of all structures presented by all possible antigens [138, 139]. Structures are non scalar objects, and there is no "structure landscape". In particular, the notion of a (local) optimum does not make sense in this context, since the structures do not admit an order.

On the other hand, we can at least define a notion of similarity between secondary structures. RNA secondary structures can be represented uniquely as rooted planar trees [140, 141, 24]. The conversion assigns an internal node to each base pair and a leaf to each unpaired digit. (An extra root is conveniently added in order to prevent the formation of a forest.) An alternative tree representation, known as homeomorphically irreducible trees, is obtained by converting connected unpaired regions and connected stacked regions, respectively, into single nodes, that can additionally be given a weight corresponding to the size of the structural element. Comparison of trees by means of edit distances is a well known technique in computer science [142, 143]. This method provides a metric distance measure between trees, and thus also between secondary structures. An alternative method for the comparison of RNA secondary structures is based on the so-called mountain representation [144]. The secondary structures are encoded as linear strings with balanced parentheses representing the base pairs, and some other symbol coding for unpaired positions. Distances are computed by direct end-to-end alignment of these strings. It has been shown that the results reported in this contribution are almost independent of the actual choice of the folding algorithm, the parameter set for the folding and the distance measures for secondary structures [33]. In this section we will use the notation $D(a, b)$ for a distance between secondary structures; thus $(\mathcal{Y}, D)$ is a metric space.

Definition 1. A map $f : \mathcal{C} \to \mathcal{Y}$ from the configuration space $\mathcal{C}$ into another metric space $\mathcal{Y}$ is called a *combinatory map*.

A landscape is special combinatory map with $\mathcal{Y} = \mathbb{R}$. The conceptual difference is that landscapes have the order of the (fitness) values as an additional feature.

Correlation in Combinatory Maps. A theory of combinatory maps analogous to the theory of landscapes discussed in the previous sections can be constructed based on based on the definition of ρ_* in sect. 4.6.1 by replacing the squared differences by squared distances in $\mathcal{Y}$. It has been proposed in [20, 24]

Definition 2. The correlation function of a combinatorial map and a "combinatorial random field", respectively, are given by

$$\hat{\rho}(\mu) = 1 - \frac{\langle D^2(f(x), f(y)) \rangle_{(x,y) \in \mu}}{\langle D^2(f(p), f(q)) \rangle}$$

$$\rho(\mu) = 1 - \frac{\mathcal{E}[D^2(f(x), f(y))]}{\mathcal{E}[D^2(f(p), f(q))]} \qquad \text{for } (x, y) \in \ddot{\mu} \text{ and } (p, q) \in v \times V$$

The fact that there are no mean values suggests that we say a combinatorial random field is isotropic if $\mathcal{E}[D^2(f(x), f(y))]$ is constant on a symmetry class μ. The notion of empirical isotropy also carries over: the squared differences are simply replaced by the corresponding squared distances in sect. 4.7. In the same spirit we may define the unscaled and scaled structure correlation lengths ℓ and ξ, respectively.

The correlation length may serve as a statistical measure of the hardness of optimization problems for a particular heuristic algorithm. The structure correlation length allows for a similar interpretation. The shorter the correlation length, the more likely is a structural change occurring as a consequence of mutation. The correlation length thus measures stability against mutation.

More detailed information on a combinatory map can be obtained from the conditional probabilities

$$P(D, \mu) = \text{Prob}\{ D(f(x), f(y)) = D \mid (x, y) \in \mu \}.$$

that the images of two randomly chosen configuration have a distance D in shape space *given* that the pair of configurations (x, y) belongs to the symmetry class μ in configuration space. We call $P(D|\mu)$ a *density surface*. The autocorrelation function can be calculated from a density surface using

$$\langle D^2(f(x), f(y)) \rangle_{(x,y) \in \mu} = \sum_D D^2 \, P(D|\ddot{\mu})$$

$$\langle D^2(f(p), f(q)) \rangle_{p,q \in V} = \sum_\mu p(\mu) \langle D^2(f(x), f(y)) \rangle_{(x,y) \in \ddot{\mu}} \,.$$

The density surfaces contain of course more information than the autocorrelation function. For instance, the probability for finding a neutral neighbor, i.e., for finding a mutant that folds into the same structure, is given by $p_{NN} = P(0|1)$.

The Combinatory Maps of RNA Secondary Structures. The number of secondary structures which are acceptable as minimum free energy structures of RNA molecules can be computed from combinatorics of structural elements [145, 29, 146, 147]. It is much smaller than the number of different sequences, since we have

$$S_n \sim 1.485 \times n^{-3/2}(1.849)^n$$

different secondary structures for κ^n sequences, i.e., there are even less structures than there are binary sequences.

The mapping from sequence space into shape space cannot be inverted: many sequences have to fold into the same secondary structure. We find that the frequency distribution of sequences folding into a common structure follows a generalized Zipf's law

$$\phi(r) = a(r + b)^{-c}$$

where r is the rank of a secondary structure S by the number of sequences folding into this particular structure and $\phi(r)$ is the fraction of sequences folding into S. The parameters b and c describe the number of very frequent structures and the fall-off of the tail of very rare structures, respectively. a is a normalization factor. The frequency distribution of structures has a very sharp peak: relatively few structures are very common, many structures are rare and play no statistically significant role.

An almost linear increase of the strucure correlation length ℓ with chain length is observed for RNA sequences. Substantial differences are found in the correlation lengths derived from different base pairing alphabets. In particular, the structures of natural **AUGC** sequences are much more stable against mutation than pure **GC**-sequences or pure **AU**-sequences.

Table 9. Asymptotic behaviour of correlation length for RNA secondary structures.

	ΔG			
Alphabet	MFE	MFE[s][+]	PF	KIN
GC	0.0222	0.0245	~ 0.02	0.0224
AU	0.0296			
GU	0.0479			
AUI	0.0299			
AGC	0.1414			
AUC	0.0792			
AUG	0.0768			
UGC	0.3135			
GCAU	0.0524	0.0414	~ 0.15	0.0616
GCAU[%]	0.0659			
GCXK	0.0491			
ABCDEF	0.0793			

[+]See table5.2.
[%]GU pairs forbidden.

It provides a plausible explanation for the use of two base pairs in nature: optimization in an RNA world with only one base pair would be very hard, and the base pairing probability in sequences with three base pairs is rather low and hence most random sequences of short chain lengths ($n < 50$) do not form thermodynamically stable structures. The choice of two base pairs thus appears to be a compromise between stability against mutation and thermodynamic stability. An alternative explanation for the usage of two base pairs in nature was published recently [148]: the current alphabet is understood to be optimal in an RNA world where replication fidelity decreases and catalytic efficiency increases with alphabet size. Both hypothesis are experimentally testable and hence we may expect a decision in favor of one of the two alternatives in the future.

In order to gain more information on the relation between RNA sequences and structures an inverse folding algorithm which determines the sequences that share the same minimum free energy secondary structure was conceived and applied to a variety of different structures [29]. One finds that sequences folding into the same secondary structure are, in essence, randomly distributed. Since there are relatively few common structures and the sequences folding into the same structure are randomly distributed in sequence space, all common structrues are found in relatively small patches of sequence space. For natural **AUGC**-sequences of chain length $n = 100$, for instance, we can expect all common structures to be found in a sphere of radius $h = 16$ in Hamming distance. There are as many as 6.2×10^{25} sequences in such a ball. Although this number is large, it is nothing compared to the total number of sequences of this chain length: $4^{100} \approx 1.6 \times 10^{60}$.

On the other hand there is a high probability for finding neutral neighbors. For the biophysical alphabet one finds, for instance, $p_{NN} \approx 0.4$ for long enough chains. Hence the sequences folding into a given target structure are not isolated in sequence space. In order to refine our understanding of their geometrical arrangement in sequence space a computer experiment was carried out allowing for an estimate of the diameter of connected sets of neutral sequences. We search for "neutral paths" through sequence space. The Hamming distance from the reference increases monotonously along such a neutral path but the structure remains unchanged. A neutral path ends when no further neutral sequence is found in the neighbourhood of the last sequence. Biophysics forces us to allow for a modification of the notion of neighborhood; we allow point mutations as well as the replacement of a base pair by another type of base pair. The length $\mathcal{L}$ of a path is the Hamming distance between the reference sequence and the last sequence, and hence a lower bound on the diameter of the connected "neutral network". Clearly, a neutral path cannot be longer than the chain length, $\mathcal{L} \leq n$. It turns out that a fraction of as many as 20% of all neutral walks have length $\mathcal{L} = n$ for **AUGC sequences** for length $n = 100$. They lead through the entire sequence space to a sequence differing in all positions from the reference but still sharing its structure. In shape spaces derived from binary sequences almost no neutral path reaches the complementary sequence. This is certainly a consequence of the symmetry of the binomial distribution: there are very few sequences in the error classes $n-1$, $n-2$, etc., and it is unlikely that we find one among them which folds precisely into the same structure as the reference sequence. Still the average length $\mathcal{L}$ is much larger than the average distance of two randomly chosen sequences, $\mathcal{L} \gg n/2$. For details see [29].

The union of all neutral paths forms a dense neutral network in the example considered here. This, of course, need not be the case in general: we may have short neutral paths confined to small disjoint regions in sequence space. We may expect a chracteristic change in the hardness of the optimization problem depending on whether the network of neutral paths is below or above a percolation threshold. It is worth noting that neutral nets are not a peculiarity of the few most frequent structures. Even the rarest structures we were able to find give rise

to networks that reach way beyond the average distance of random sequences
[30].

Random combinatory maps with prescribed correlation structure are not as
easy to construct as random fields. As a first step one considers only the pre-
image of a certain structure, i.e., the set of all sequences folding into the same
secondary structure. This set forms a sub-graph of the configuration space. The
structure of such random sub-graphs with a prescribed probability for neutral
neighbors, that is, with a prescribed vertex degree has been investigated in detail
[75]. In particular, explicit expressions for the fraction of neutral mutations such
that the subgraph is dense and connected have been obtained for Hamming
graphs. A random sub-graph of a Hamming graph with a alphabet of size α is
dense and connected in the limit of large systemsize n if and only if

$$p_{NN} \geq 1 - \sqrt[\alpha - 1]{\alpha^{-1}}.$$

6 Conclusions

The basic ingredient of the theory of landscapes presented in this contribution is
the notion of Fourier series defined on arbitrary graphs. The detailed mathematical form of these series depends crucially on the geometry of the configuration
space on which the landscape, or a random field model of a landscape, is defined
in that the basis functions are eigenvectors of the graphs Laplacian. Landscapes
which are such eigenvectors are termed elementary as they play a special role in
the theory. Surprisingly, we find that almost all the well known combinatorial
optimization problems (with the exception of the asymmetric TSP) are elementary. The main subject of this contribution is therefore a theory of elementary
landscapes and random fields.

It turns out the correlation structure of landscapes and random fields is
closely linked to these generalized Fourier series, and therefore to the spectrum
of the graph Laplacian. In particular the geometry of local optima can be linked
to the graph Laplacian. Isotropic random fields are characterized in terms of their
Fourier coefficients. All landscapes on the Boolean hypercube can be represented
as a superposition of p-spin Hamiltonians, and isotropic Gaussian landscapes
correspond exactly to a superposition of Derrida's version of p-spin models:
the coupling coefficients are uncorrelated and the variance of the coefficients
depends only on the order p of the coupling. The Fourier expansion of (fitness)
landscapes can be used to find the set of all possible correlation functions; more
precisely, the set of all autocorrelation functions forms a simplex spanned by
suitably normalized left eigenvectors of the collapsed adjacency matrix of the
configuration space. The corresponding eigenvalue determines the decay of the
correlation function along a simple random walk on the landscape.

The formalism derived in this contribution suggests to approximate a given
landscape by a superposition of statistical models that are elementary landscapes
(which are hopefully easier to understand). The coefficients of such a superposition can be obtained by observing that the correlation function is a superposition
of the correlation functions of the elementary landscapes under consideration.

The latter can be computed analytically for a number of problems, since they are the left eigenvectors of the collapsed adjacency matrix of the underlying configuration space. Instead of the correlation function we can then use the coefficients of this decomposition as a means of characterizing a landscape. On a Boolean hypercube, for instance, this spectrum can be interpreted directly as the relative importance of p-ary (spin) interactions.

As a biological application we have studied RNA free energy landscapes. It turns out that the natural **GCAU** landscape can be fitted reasonably well by a superposition of a 1-spin and a 6-spin model with equal amplitude, while the **GC** and **GCXK** landscapes can be approximated by a 6-spin model. We do not expect that RNA landscapes are elementary, we rather expect a distribution of p-spin models centered around $p = 6$. The numerical correlation data are to noisy, however, to permit a higher resolution, and we have to be content with the approximate location of the peak.

An extension of the notion of fitness landscapes to general sequence structure relations is outlined very briefly in the last sections. Based on an alternative form the autocorrelation one defines a generalized correlation function for mappings from one metric space into another one. Whether an anlogy of the Fourier theory of landscapes is possible also for such "combinatory maps" and their random field models is unknown at present.

Acknowledgements

Fruitful discussions with Andreas Dress, Walter Fontana, Ivo Hofacker, Stuart Kauffman, Christian Reidys, Peter Schuster, and Günter Wagner are gratefully acknowledged.

References

1. Sewall Wright. The roles of mutation, inbreeding, crossbreeeding and selection in evolution. In D. F. Jones, editor, *Int. Proceedings of the Sixth International Congress on Genetics*, pages 356–366, 1932.

2. K. Binder and A. P. Young. Spin glasses: experimental facts, theoretical concepts, and open questions. *Rev.Mod.Phys.*, 58:801–976, 1986.

3. M. Mézard, G. Parisi, and M.A. Virasoro. *Spin Glass Theory and Beyond*. World Scientific, Singapore, 1987.

4. M.R. Garey and D.S. Johnson. *Computers and Intractability. A Guide to the Theory of $\mathcal{NP}$ Completeness*. Freeman, San Francisco, 1979.

5. M. Eigen. Selforganization of matter and the evolution of biological macro-molecules. *Die Naturwissenschaften*, 10:465–523, 1971.

6. W. Fontana and P. Schuster. A computer model of evolutionary optimization. *Biophysical Chemistry*, 26:123–147, 1987.

7. Walter Fontana, Wolfgang Schnabl, and Peter Schuster. Physical aspects of evolutionary optimization and adaption. *Physical Review A*, 40(6):3301–3321, 1989.

8. C. Amitrano, L. Peliti, and M. Saber. Population dynamics in a spin-glass model of chemical evolution. *J. Mol. Evol.*, 29:513–525, 1989.

9. M. Eigen, J. McCaskill, and P. Schuster. The molecular Quasispecies. *Adv. Chem. Phys.*, 75:149 – 263, 1989.

10. Peter F. Stadler and Wolfgang Schnabl. The landscape of the traveling salesman problem. *Phys. Letters A*, 161:337–344, 1992.

11. Peter F. Stadler. Correlation in landscapes of combinatorial optimization problems. *Europhys. Lett.*, 20:479–482, 1992.

12. Peter F. Stadler and Robert Happel. Correlation structure of the landscape of the graph-bipartitioning-problem. *J. Phys. A.: Math. Gen.*, 25:3103–3110, 1992.

13. Catherine A. Macken and Alan S. Perelson. Protein evolution on rugged landscapes. *Proc.Natl.Acad.Sci.USA*, 86:6191–6195, 1989.

14. C.A. Macken, P.S. Hagan, and A.S. Perelson. Evolutionary walks on rugged landscapes. *SIAM J.Appl.Math.*, 51:799–827, 1991.

15. Henrik Flyvbjerg and Benny Lautrup. Evolution in a rugged fitness landscape. *Phys. Rev. A*, 46:6714–6723, 1992.

16. P. Bak, H. Flyvbjerg, and B. Lautrup. Coevolution in a rugged fitness landscape. *Phys.Rev. A[15]*, 46:6724–6730, 1992.

17. S. A. Kauffman and S. Levin. Towards a general theory of adaptive walks on rugged landscapes. *J. Theor. Biol.*, 128:11, 1987.

18. S. A. Kauffman and E. D. Weinberger. The n-k model of rugged fitness landscapes and its application to maturation of the immune response. *J. Theor. Biol.*, 141:211, 1989.

19. Edward D. Weinberger. Local properties of Kauffman's N-k model: a tunably rugged energy landscape. *Phys. Rev. A*, 44(10):6399–6413, 1991.

20. Walter Fontana, Peter F. Stadler, Erich G. Bornberg-Bauer, Thomas Griesmacher, Ivo L. Hofacker, Manfred Tacker, Pedro Tarazona, Edward D. Weinberger, and Peter Schuster. RNA folding and combinatory landscapes. *Phys. Rev. E*, 47(3):2083 – 2099, 1993.

21. E.D. Weinberger and P.F. Stadler. Why some fitness landscapes are fractal. *J. Theor. Biol.*, 163:255–275, 1993.

22. Sebastian Bonhoeffer and Peter F. Stadler. Errortreshold on complex fitness landscapes. *J. Theor. Biol.*, 164:359–372, 1993.

23. W. Fontana, T. Griesmacher, W. Schnabl, P.F. Stadler, and P. Schuster. Statistics of landscapes based on free energies, replication and degradation rate constants of RNA secondary structures. *Monatsh. Chemie*, 122:795–819, 1991.

24. W. Fontana, D. A. M. Konings, P. F. Stadler, and P. Schuster. Statistics of rna secondary structures. *Biochemistry*, 33:1389–1404, 1993.

25. Ivo L. Hofacker, Walter Fontana, Peter F. Stadler, Sebastian Bonhoeffer, Manfred Tacker, and Peter Schuster. Fast folding and comparison of RNA secondary structures. *Monatsh. Chemie*, 125(2):167–188, 1994.

26. M.A. Huynen and P. Hogeweg. Pattern generation in molecular evolution. Exploitation of the variation in RNA landscapes. *J.Mol.Evol.*, 39:71–79, 1994.

27. Martijn A. Huynen, Peter F. Stadler, and Walter Fontana. Evolution of RNA and the Neutral Theory. 1995. SFI Preprint # 95-01-006.

28. P. Schuster. Complex optimization in an artificial RNA world. In D. Farmer, C. Langton, S. Rasmussen, and C. Taylor, editors, *Artificial Life II*, pages 277–291, Addison-Wesley, 1992.

29. Peter Schuster, Walter Fontana, Peter F Stadler, and Ivo L Hofacker. From sequences to shapes and back: a case study in RNA secondary structures. *Proc.Roy.Soc.Lond.B*, 255:279–284, 1994.

30. Peter Schuster and Peter F Stadler. Landscapes: complex optimization problems and biopolymer structures. *Computers Chem.*, 18:295–314, 1994.

31. Peter F. Stadler and Walter Grüner. Anisotropy in fitness landscapes. *J. Theor. Biol.*, 165:373–388, 1993.

32. Manfred Tacker, Walter Fontana, Peter Stadler, and Peter Schuster. Statistics of RNA melting kinetics. *Eur. J. Biophys.*, 23:29–38, 1994.

33. Manfred Tacker, Peter F. Stadler, Erich G. Bornberg-Bauer, Ivo L. Hofacker, and Peter Schuster. Robust properties of RNA secondary structure folding algorithms. 1005. In preparation.

34. E. L. Lawler, J. K. Lenstra, A. H. G Rinnoy Kan, and D. B. Shmoys. *The Traveling Salesman Problem. A Guided Tour of Combinatorial Optimization.* John Wiley & Sons, 1985.

35. R. G. Bland and D. F. Shallcross. Large traveling salesmen problems arising from experiments in x-ray crystallography. *Oper.Res.Lett.*, 8:125–128, 1988.

36. D. Chan and D. Mercier. IC insertion: an application of the TSP. *Int.J.Prod.Res.*, 3:9–28, 1989.

37. H. Bohr and S. Brunak. Travelling salesman approach to protein conformation. *Complex Systems*, 3:9–28, 1990.

38. D.L. Miller and J. F. Pekny. Exact solution of large asymmetric traveling salesman problems. *Science*, 251:754–761, 1991.

39. S. Lin and B.W. Kernighan. An effective heuristic algorithm for the traveling salesman problem. *Oper.Res.*, 21:498–516, 1965.

40. H. Wielandt. *Finite Permutation Groups.* Academic Press, New York, 1964.

41. D. G. Higman. Intersection matrices for finite permutation groups. *J. Algebra*, 6:22–42, 1967.

42. Norman Biggs. *Finite Groups of Automorphisms.* Volume 6 of *London Mathematical Society Lecture Notes*, Cambridge University Press, Cambridge UK, 1971.

43. Bela Bollobás. *Graph Theory – An Introductory Course.* Springer-Verlag, New York, 1979.

44. C. D. Godsil. *Algebraic Combinatorics.* Chapman & Hall, New York, 1993.

45. P. Delsarte. *An algebraic approach to association schemes of coding theory.* Volume 10 of *Phillips Research Reports Supplements*, Phillips, 1973.

46. Norman L. Biggs. *Algebraic Graph Theory.* Cambridge University Press, Cambridge UK, 2nd edition, 1994.

47. F. Spitzer. *Principles of Random Walks.* Springer-Verlag, New York, 1976.

48. D.M. Cvetković, M. Doob, and H. Sachs. *Spectra of Graphs – Theory and Applications.* Academic Press, New York, 1980.

49. Peter F. Stadler and Robert Happel. Canonical approximation of landscapes. 1994. Submitted to J.Stat.Phys.

50. Paolo M. Soardi. *Potential Theory on Infinite Networks.* Volume 1590 of *Lecture Notes in Mathematics*, Springer-Verlag, Berlin, 1994.

51. Mark Kac. Can you hear the shape of a drum. *Am.Math.Monthly*, 73(4):1–23, 1966.

52. G.A. Baker. Drum shapes and isospectral graphs. *J.Math.Phys.*, 7:2238, 1966.

53. M.E. Fisher. On hearing the shape of a drum. *J.Comb.Theory*, 1:105–125, 1966.

54. C.T. Benson and J.B. Jacobs. On hearing the shape of combinatorial drums. *J.Comb.Theory(B)*, 13:170–178, 1972.

55. Gert Sabidussi. Vertex transitive graphs. *Mh. Math*, 68:426–438, 1964.

56. Norman Biggs. *Algebraic Graph Theory*. Cambridge University Press, Cambridge UK, 1st edition, 1974.

57. Peter F. Stadler. Random walks and orthogonal functions associated with highly symmetric graphs. *Disc. Math.*, 1994. in press.

58. G.M. Adel'son-Velskii *et al.* Example of a graph without a transitive automorphism group. *Soviet. Math. Dokl.*, 10:440–441, 1969. Russian.

59. Edward D. Weinberger. Fourier and Taylor series on fitness landscapes. *Biological Cybernetics*, 65:321–330, 1991.

60. L. Lovász. Spectra of graphs with transitive groups. *Periodica Math.Hung.*, 6:191–195, 1975.

61. J.-P. Serre. *Linear Representations of Finite Groups*. Springer-Verlag, New York, Heidelberg, Berlin, 1977.

62. A.J. Schwenk. Computing the characteristic polynomial of a graph. In *Graphs and Combinatorics*, pages 153–162, Springer-Verlag, Berlin, 1974.

63. D.L. Powers and M.M. Sulaiman. The walk partition and colorations of a graph. *Linear Algebra Appl.*, 48:145–159, 1982.

64. D.L. Powers. Eigenvectors of distance-regular graphs. *SIAM J. Matrix Anal. Appl.*, 9:399–407, 1988.

65. D.M. Cvetković, M. Doob, and H. Sachs. *Spectra of Graphs – Theory and Applications*. Volume New York, Academic Press, 1980.

66. R. W. Hamming. Error detecting and error correcting codes. *Bell Syst.Tech.J.*, 29:147–160, 1950.

67. A.W.M. Dress and D.S. Rumschitzki. Evolution on sequence space and tensor products of representation spaces. *Acta Appl.Math.*, 11:103–111, 1988.

68. A.E. Brouwer, A.M. Cohen, and A. Neumaier. *Distance-regular Graphs*. Springer Verlag, Berlin, New York, 1989.

69. C.F. Dunkl. A Krawtchouk polynomial addition theorem and wreath products of symmetric groups. *Indiana Univ.Math.J.*, 25:335–358, 1976.

70. T.H. Koornwinder. Krawtchouk polynomials. A unification of two different group theoretic interpretations. *SIAM J.Math.Anal.*, 13:1011–1023, 1982.

71. M. Krawtchouk. Sur une gènèralisation des polynomes d'Hermite. *Comptes Rendus*, 189:620–622, 1929.

72. F.J. MacWilliams and N.J.A. Sloane. *The Theory of Error-Correcting Codes*. North-Holland, Amsterdam, New York, Oxford, Tokyo, 1991.

73. D. Rumschitzky. Spectral properties of eigen's evolution matrices. *J.Math.Biol.*, 24:667–680, 1987.

74. J.H. vanLint. *Introduction to Coding Theory*. Springer-Verlag, New York, 1982.

75. Christian Reidys, Peter Schuster, and Peter F Stadler. Generic properties of combinatory maps and application on RNA secondary structures. 1995. Preprint.

76. Christian Reidys and Christian Forst. Replication on neutral networks in rna induced by rna secondary structures. 1994. Preprint.

77. P.J. Cameron and J.H. vanLint. *Designs, Graphs, Codes, and their Links*. Volume 22 of *London Math. Soc. Student Texts*, Cambridge University Press, Cambridge UK, 1991.

78. C.F. Dunkl. Orthogonal functions on some permutation groups. In D.K. Ray-Chaudhuri, editor, *Proceeding of Symposia in Pure Mathematic Vol. 34*, American Mathematical Society, New York, 1979.

79. Normal L. Biggs and A.T. White. *Permutation Groups and Combinatorial Structures*. Cambridge University Press, Cambridge UK, 1979.

80. P. Diaconis and M. Shahshahani. Generating a random permutation with random transpositions. *Z. Wahrscheinlichkeitsth. verw. Gebiete*, 57:159–179, 1981.

81. R.E. Ingram. Some characters of the symmetric group. *Proc.Amer.Math.Soc.*, 1:358–369, 1950.

82. I.G. MacDonald. *Symmetric Functions and Hall Polynomials*. Oxford Univ. Press, Oxford UK, 1979.

83. Julian Besag. Spatial interactions and the statistical analysis of lattice systems. *Amer. Math. Monthly*, 81:192–236, 1974.

84. M.J.E. Golay. Sieves for low-autocorrelation binary sequences. *IEEE Trans. Inform. Th.*, IT-23:43–51, 1977.

85. H. Hotelling. Analysis of a complex of statistical variables into principal components. *J.Educ.Psych.*, 24:417–441 and 498–520, 1933.

86. C.R. Rao. *Linear Statistical Interference and Its Applications*. Wiley, New York, 2nd edition, 1973.

87. M.S. Bartlett. *An Introduction to Stochastic Processes*. Cambridge University Press, Cambridge UK, 1955.

88. P. Whittle. Stochastic processes in several dimensions. *Bull. Int. Statist. Inst.*, 40:974–994, 1963.

89. G.R. Grimmet. A theorem about random fields. *Bull. London Math. Soc.*, 5:81–85, 1973.

90. John Moussouris. Gibbs and Markov systems with constraints. *J. Stat. Phys.*, 10:11–33, 1974.

91. M.B. Averintsev. On a method of describing complete parameter fields. *Problemy Peredaci Informatsii*, 6:100–109, 1970.

92. R.L. Dobrushin. The description of a random field by means of its conditional probabilities, and conditions of its regularities. *Th. Prob. & Appl.*, 13:197–224, 1968.

93. Frank Spitzer. Markov random fields and gibbs ensembles. *Amer. Math. Monthly*, 78:142–154, 1971.

94. S. Karlin and H.M. Taylor. *A first course in stochastic processes*. Academic Press, New York, 1975.

95. Peter F. Stadler. Linear operators on correlated landscapes. *J.Physique*, 4:681–696, 1994.

96. B. Derrida. Random energy model: limit of a family of disordered models. *Phys.Rev.Lett.*, 45:79–82, 1980.

97. David Sherrington and Scott Kirkpatrick. Solvable model of a spin-glass. *Physical Review Letters*, 35(26):1792 – 1795, 1975.

98. B. Derrida. The random energy model. *Phys.Rep.*, 67:29–35, 1980.

99. Bernard Derrida. Random-energy model: an exactly solvable model of disorderes systems. *Phys. Rev. B*, 24(5):2613–2626, 1981.

100. E. Gardner and B. Derrida. The probability distribution of the partition function of the random energy model. *J.Phys. A*, 22:1975–1982, 1989.

101. Edward D. Weinberger. Correlated and uncorrelated fitness landscapes and how to tell the difference. *Biol.Cybern.*, 63:325–336, 1990.

102. C.W. Gardiner. *Handbook of Stochastic Methods*. Springer-Verlag, Berlin, 2nd edition, 1990.

103. S.A. Kauffman. *The Origin of Order*. Oxford University Press, New York, Oxford, 1993.

104. K.J. Laidler. *Chemical Kinetics*. Harper, New York, 3rd edition, 1992.

105. L.K. Grover. Local search and the local structure of NP-complete problems. *Oper.Res.Lett.*, 12:235–243, 1992.

106. W. Miller Jr. *Symmetry and Separation of Variable.* Volume 4 of *Enceyclopedia of Mathematics and its Applications,* Cambridge Univ. Press, Cambridge, UK, 1984.

107. R.A. Brualdi and H.J. Ryser. *Combinatorial Matrix Theory.* Cambridge Univ. Press, Cambridge UK, 1991.

108. I. Chavel. *Eigenvalues in Riemannian Geometry.* Academic Press, Orlando Fl., 1984.

109. R. Palmer. Optimization on rugged landscapes. In A. S. Perelson and S. A. Kauffman, editors, *Molecular Evolution on Rugged Landscapes: Proteins, RNA, and the Immune System,* pages 3–25, Addison Wesley, Redwood City, CA, 1991.

110. A.J. Bray and M.A. Moore. Metastable states in spin glasses. *J.Phys.C:Solid St.Phys.*, 13:L469–L476, 1980.

111. A.J. Bray and M.A. Moore. Metastable states in spin glasses with short-ranged interactions. *J.Phys.C:Solid St.Phys.*, 14:1313–1327, 1981.

112. C. De Dominicis, M. Gabay, T. Garel, and H. Orland. White and weighted averages over solutions of the Thouless Anderson Palmer equations for the Sherrington Kirkpatrick spin glass. *J.Physique*, 41:923–930, 1980.

113. Bernard Derrida and E. Gardner. Metastable states of a spin glass chain at 0 temperature. *J.Physique*, 47:959–965, 1986.

114. F. Tanaka and S.F. Edwards. Analytic theory of the ground state properties of a spin glass: I. Ising spin glass. *J.Phys.F:Metal Phys.*, 10:2769–2778, 1980.

115. Alan S. Perelson and Catherine A. Macken. Protein evolution on partially correlated landscapes. Santa Fe Institute Preprint 94-11-060.

116. D.J. Thouless, P.W. Anderson, and R.G. Palmer. *Phil.Mag.*, 35:593, 1977.

117. Catherine A. Macken and Peter F. Stadler. Rugged landscapes. 1995. To appear in SFI summerschool volume 1993.

118. E.H.L. Aarts and J. Korst. *Simulated Annealing and Boltzman Machines.* J. Wiley & Sons, New York, 1990.

119. R.H.J.M. Otten and L.P.P.P. vanGinneken. *The Annealing Algorith.* Kluwer Acad.Publ., Boston, 1989.

120. Y. Fu and P. W. Anderson. Application of statistical mechanics to NP-complete problems in combinatorial optimization. *J.Phys.A:Math.Gen.*, 19:1605–1620, 1986.

121. J. Bernasconi. Low autocorrelation binary sequences: statistical mechanics and configuration space analysis. *J.Physique*, 48:559–567, 1987.

122. Bärbel Krakhofer. *Local Optima in Landscapes of Combinatorial Optimization Problems.* Master's thesis, University of Vienna, Dept. of Theoretical Chemistry, 1995.

123. G. B. Sorkin. *Combinatorial optimization, simulated annealing, and fractals.* Technical Report RC13674 (No.61253), IBM Research Report, 1988.

124. R. Voss. Characterization and measurement of random fractals. *Physical Scripta*, T13:257–260, 1986.

125. M. Mézard and G. Parisi. Replicas and optimization. *J.Physique Lett.*, 46:L771–L778, 1986.

126. M. Zuker. The use of dynamic programming algorithms in RNA secondary structure prediction. In Michael S. Waterman, editor, *Mathematical Methods for DNA Sequences,* pages 159–184, CRC Press, 1989.

127. M. Zuker and D. Sankoff. RNA secondary structures and their prediction. *Bull.Math.Biol.*, 46(4):591–621, 1984.

128. H. M. Martinez. An RNA folding rule. *Nucl.Acid.Res.*, 12:323–335, 1984.

129. Manfred Tacker. *Robust Properties of RNA Secondary Structure Folding Algorithms.* PhD thesis, University of Vienna, 1993.

130. John S. McCaskill. The equilibrium partition function and base pair binding probabilities for RNA secondary structure. *Biopolymers*, 29:1105–1119, 1990.

131. Ruth Nussinov, George Piecznik, Jerrold R. Griggs, and Daniel J. Kleitman. Algorithms for loop matching. *SIAM J. Appl. Math.*, 35(1):68–82, 1978.

132. A. A. Mironov, L. P. Dyakonova, and A. E. Kister. A kinetic approach to the prediction of RNA secondary structures. *Journal of Biomolecular Structure and Dynamics*, 2(5):953, 1985.

133. A. A. Mironov and A. E. Kister. RNA secondary structure formation during transcription. *J. of Biomolecular Structure and Dynamics*, 4:1–9, 1986.

134. M. Zuker. `mfold-2.0`. `pub/mfold.tar.Z` `nrcbsa.bio.nrc.ca`. (Public Domain Software).

135. I. L. Hofacker, W. Fontana, P. F. Stadler, L. S. Bonhoeffer, M. Tacker, and P. Schuster. `Vienna RNA Package`. `pub/RNA/ViennaRNA-1.03` `ftp.itc.univie.ac.at`. (Public Domain Software).

136. W. Salser. Globin messenger RNA sequences — analysis of base-pairing and evolutionary implications. *Cold Spring Harbour Symp. Quant. Biol.*, 42:985, 1977.

137. Susan M. Freier, Ryszard Kierzek, John A. Jaeger, Naoki Sugimoto, Marvin H. Caruthers, Thomas Neilson, and Douglas H. Turner. Improved free-energy parameters for predictions of RNA duplex stability. *Proc. Natl. Acad. Sci., USA*, 83:9373–9377, 1986.

138. A. S. Perelson and G. Oster. Theoretical studies of clonal selection: minimal antibody repertoire size and reliability of self/ non-self discrimination. *Journal of Theoretical Biology*, 81:645–670, 1979.

139. L. A. Segel and A. P. Perelson. Computations in shape space: a new approach to immune network theory. In *Theoretical Immunology. Part Two*, pages 321 – 343, Addison-Wesley, Redwood City (Cal.), 1988.

140. Bruce A. Shapiro. An algorithm for comparing multiple RNA secondary stuctures. *CABIOS*, 4(3):387–393, 1988.

141. Bruce A. Shapiro and Khaizhong Zhang. Comparing multiple RNA secondary structures using tree comparisons. *CABIOS*, 6:309–318, 1990.

142. K. Tai. The tree-to-tree correction problem. *J. ACM*, 26:422–433, 1979.

143. K. Ohmori and E. Tanaka. A unified view on tree metrics. In G. Ferrate, editor, *Syntactic and Structural Pattern Recognition*, pages 85–100, Springer-Verlag, Berlin, Heidelberg, 1988.

144. Pauline Hogeweg and B. Hesper. Energy directed folding of RNA sequences. *Nucleic acids research*, 12:67–74, 1984.

145. I.L. Hofacker, P. Schuster, and P.F. Stadler. Combinatorics of secondary structures. *submitted to SIAM J. Disc. Math.*, 1993.

146. P.R. Stein and M.S. Waterman. On some new sequences generalizing the Catalan and Motzkin numbers. *Discrete Mathematics*, 26:261–272, 1978.

147. M. S. Waterman. Secondary structure of single - stranded nucleic acids. *Studies on foundations and combinatorics, Advances in mathematics supplementary studies, Academic Press N.Y.*, 1:167 – 212, 1978.

148. E. Szathmáry. Four letters in the genetic alphabet: a frozen evolutionary optimum? *Proc.Roy.Soc. London B*, 245:91–99, 1991.

Part V

Coarsening Phenomena in One Dimension

Coarsening Phenomena in One Dimension

B. Derrida

Laboratoire de Physique Statistique, ENS
24 rue Lhomond, 75005 Paris, France
and
Service de Physique Théorique, CE
Saclay, F91191 Gif sur Yvette, France

1 Introduction

Coarsening phenomena occur in a large variety of situations in physical systems: spinodal decomposition [1], grain growth [2, 3, 4], soap froths [5, 6], breath figures [7, 8, 9]. The purpose of this lecture is to describe effects related to coarsening in the dynamics of systems as simple as the one dimensional Ising or Potts model at zero temperature [10, 11, 12, 9, 13, 14, 15]. When the initial condition is random, these systems exhibit, in the long time limit, a scaling regime where the size of domains grows like $t^{1/2}$. Several approaches (analogy with random walks and reaction diffusion models) can be used to understand this regime. These approaches as well as some approximations(finite size scaling, mean field theory) are discussed in what follows.

2 Finite temperature dynamics of the Ising chain

The equilibrium properties of the one dimensional Ising model have been understood for a long time: the calculation of the partition function or of any correlation function at equilibrium is elementary. Let us discuss here the dynamics of the Ising chain.

Consider a chain of N Ising spins ($S_i = \pm 1$) on a ring evolving according to the Glauber dynamics [16]. During any time interval dt (with $dt \ll 1$), each spin i in the system has a probability dt of being updated; when it is updated, it becomes $+1$ with probability $[1 + \tanh(\frac{S_{i+1}+S_{i-1}}{T})]/2$ or -1 with probability $[1 - \tanh(\frac{S_{i+1}+S_{i-1}}{T})]/2$. Here T is the temperature and the periodic boundary conditions imply that $S_0 = S_N$ and $S_{N+1} = S_1$. This process is stochastic and if one denotes by $\langle . \rangle$ the average over the process, it follows from the very definition of the dynamics that

$$\frac{d\langle S_i \rangle}{dt} = -\langle S_i \rangle + \langle \tanh \left(\frac{S_{i+1} + S_{i-1}}{T} \right) \rangle . \tag{1}$$

The simplicity of the one dimensional (and zero magnetic field) problem is due to the following identity

$$\tanh\left(\frac{S_{i+1} + S_{i-1}}{T}\right) = \frac{S_{i+1} + S_{i-1}}{2} \tanh\left(\frac{2}{T}\right) \tag{2}$$

which allows one to rewrite (1) as

$$\frac{d\langle S_i \rangle}{dt} = -\langle S_i \rangle + (\langle S_{i+1} \rangle + \langle S_{i-1} \rangle) \frac{a}{2} \tag{3}$$

where

$$a = \tanh(2/T) . \tag{4}$$

The fact that the time evolution of $\langle S_i \rangle$ involves other one point functions $\langle S_j \rangle$ but no higher correlations (like $\langle S_j S_k \rangle$) makes the problem simple enough to allow for an exact solution [10, 11, 12, 13, 9]. If initially, the system is in a given configuration $\{S_i(0)\}$, the solution of (3) can be written as

$$\langle S_i(t) \rangle = \frac{1}{2\pi} \int_0^{2\pi} d\theta \sum_{j=-\infty}^{\infty} S_j(0) \cos[(j-i)\theta] e^{-t(1-a\cos\theta)} . \tag{5}$$

Similarly, for $i \neq j$, one can write the time evolution of the two point function

$$\frac{d\langle S_i S_j \rangle}{dt} = -2\langle S_i S_j \rangle + (\langle S_{i+1} S_j \rangle + \langle S_{i-1} S_j \rangle + \langle S_i S_{j-1} \rangle + \langle S_i S_{j+1} \rangle) \frac{a}{2} . \tag{6}$$

Obviously, for $i = j$, one has $\langle S_i^2 \rangle = 1$.

There are several dynamical properties which can be understood as consequences of the evolution equations (1–6):

1. First, if initially $\langle S_i(0) \rangle = m(0)$, independent of i, one finds from (3) that

$$\langle S_i(t) \rangle = m(t) = m(0) e^{-t(1-a)} . \tag{7}$$

   So the magnetization relaxes to 0 as it should since the magnetization at equlibrium is zero, with the exception of zero temperature ($a = 1$) where the magnetization remains unchanged.

2. One can also calculate the autocorrelation function. Assume that initially, one starts with a situation where the spins are uncorrelated, that is

$$\langle S_i(0) S_j(0) \rangle = 0 .$$

   Then from (5), it follows that

$$\langle S_i(t) S_i(0) \rangle = \frac{1}{2\pi} \int_0^{2\pi} d\theta \, e^{-t(1-a\cos\theta)} . \tag{8}$$

   In the long time limit, the integral is dominated by θ close to zero and therefore

$$\langle S_i(t) S_i(0) \rangle \simeq \frac{e^{-(1-a)t}}{\sqrt{2\pi a t}} . \tag{9}$$

   The autocorrelation function decays exponentially at non zero temperature ($a < 1$) and as a power law at zero temperature.

3. A third property which can be obtained form (6) is the equal time two point function. If initially, one starts with a given correlation function

$$\langle S_i(0)S_j(0)\rangle\rangle = g(i - j) ,\tag{10}$$

the solution of (6) is for $i < j$

$$\langle S_i(t)S_j(t)\rangle = \frac{1}{\pi}\int_0^{2\pi} d\theta \; \sin[(j - i)\theta] \; e^{-t(2-2a\cos\theta)} \sum_{n=1}^{\infty} g(n) \; \sin(n\theta)$$

$$+ \frac{a}{2\pi}\int_0^{2\pi} d\theta \; \sin[(j - i)\theta] \; \sin\theta \; \frac{1 - e^{-t(2-2a\cos\theta)}}{1 - a\cos\theta} .\tag{11}$$

In the long time limit, and at non zero temperature (i.e. $a < 1$), this becomes for $j > i$

$$\langle S_i(t)S_j(t)\rangle \to \frac{a}{2\pi}\int_0^{2\pi} d\theta \; \frac{\sin[(j - i)\theta] \; \sin\theta}{1 - a\cos\theta}$$

$$= \left(\frac{1 - \sqrt{1 - a^2}}{a}\right)^{j-i} = \left[\tanh(\frac{1}{T})\right]^{j-i}\tag{12}$$

which means that the system relaxes to thermal equilibrium. Using the fact that $2\sum_{y\geq 1}[\tanh(1/T)]^y \sin(y\theta) = a\sin\theta/(1 - a\cos\theta)$, it is also easy to check from (11) that a system, initially at thermal equilibrium (i.e. such that $\langle S_i(0)S_j(0)\rangle = g(j - i) = \tanh^{j-i}(1/T)$), remains at equilibrium at any later time.

3 Zero temperature dynamics of the Ising chain

The zero temperature dynamical properties of the Ising chain follow directly from what has been done in the previous section by imposing $a = 1$.

1. First, if initially one starts with a magnetized configuration ($\langle S_i(0)\rangle = m(0)$), the magnetization does not change with time (7)

$$m(t) = \langle S_i(t)\rangle = \langle S_i(0)\rangle = m(0) .\tag{13}$$

2. Second, for an initial condition where the spins are uncorrelated, ($g(n) = 0$ for $n \geq 1$), equation (11) becomes for $j > i$

$$\langle S_i(t)S_j(t)\rangle = \frac{1}{2\pi}\int_0^{2\pi} d\theta \; \sin[(j - i)\theta] \; \sin\theta \; \frac{1 - e^{-t(2-2\cos\theta)}}{1 - \cos\theta} .\tag{14}$$

The long time limit of this expression is dominated by $\theta = O(\sqrt{t})$ and (by calculating the long time behavior of $d\langle S_i(t)S_j(t)\rangle/dt$) it is easy to show that for large t,

$$\langle S_i(t)S_j(t)\rangle \simeq f\left(\frac{|j - i|}{\sqrt{t}}\right)\tag{15}$$

where

$$f(x) = \frac{1}{\sqrt{\pi}} \int_x^\infty e^{-u^2/4} \, du \; . \tag{16}$$

We see that there is a characteristic lenght $L(t) = \sqrt{t}$ and that, as $t \to \infty$, the system reaches a scaling regime where the correlation function becomes a function of the single variable $|j - i|/L(t)$.

A particular case which will be used later is the nearest neighbor correlation function

$$\langle S_i(t) S_{i+1}(t) \rangle \simeq 1 - \frac{1}{\sqrt{\pi t}} + O\left(\frac{1}{t^{3/2}}\right) \; . \tag{17}$$

The calculation can be easily repeated for correlated initial conditions (for example $g(n)$ decaying exponentially with n). The result (15,16) remains the same and the correlations in the initial condition only affect corrections to scaling.

One property which can be calculated from the nearest neighbor correlation function is the average number of times $\langle \nu_i(t) \rangle$ that spin i has flipped up to time t

$$\langle \nu_i(t) \rangle = \int_0^t dt' \, \frac{2 - \langle [S_{i-1}(t') + S_{i+1}(t')] S_i(t') \rangle}{4} \simeq \sqrt{\frac{t}{\pi}} \; . \tag{18}$$

3. We have seen (9) that for an uncorrelated initial condition, the autocorrelation function is given, in the long time limit, by

$$\langle S_i(t) S_i(0) \rangle \simeq \frac{1}{\sqrt{2\pi t}} \; . \tag{19}$$

For large t, one can also show that (5) becomes

$$\langle S_i(t) \rangle = \frac{1}{\sqrt{2\pi t}} \sum_{j=-\infty}^{\infty} e^{-(j-i)^2/2t} \, S_j(0) \tag{20}$$

This, together with (15) leads for $t' > t$ (and when both $t \gg 1$ and $t'-t \gg 1$) to

$$\langle S_i(t') S_i(t) \rangle \simeq \frac{1}{\sqrt{2\pi(t'-t)}} \sum_{j=-\infty}^{\infty} e^{-(j-i)^2/2(t'-t)} f(|j-i|/\sqrt{t}) \tag{21}$$

which gives using (16)

$$\langle S_i(t') S_i(t) \rangle \simeq 1 - \frac{2}{\pi} \tan^{-1} \sqrt{\frac{t'-t}{2t}} \; . \tag{22}$$

We see that for large t, $\langle S_i(t') S_i(t) \rangle$ is not a function of $t' - t$, as it would, if the system was reaching an equilibrium. Instead, it is a function of t'/t which is often found in systems perpetually out of equlibrium [17, 18]. This is easy to understand because for an infinite system, the evolution never stops: the size of the domains grows for ever. The fact that the correlation function is a function of t'/t is often called aging.

Remark 1: expression (15) is a general property of coarsening phenomena which occur when an Ising model is quenched below its transition temperature. In all dimensions, for Glauber dynamics, the characteristic domain size $L(t)$ grows like $t^{1/2}$. The long time limit is a scaling regime where the system looks statistically the same at two different times t and t' up to a rescaling by a factor $L(t')/L(t)$.

Remark 2: instead of the Glauber dynamics, one can use a dynamical rule which conserves locally the order parameter (to describe situations like the phase separation of a fluid). In this case too, when the system is quenched into its two phase region, one observes a scaling regime for the growth of domains. The main difference is that the characteristic domain size $L(t) \sim t^{1/3}$ in all dimensions instead of the $t^{1/2}$ characteristic of the non-conserved case [1].

Remark 3: what is special to the one dimensional case is that the low temperature phase is limited to $T = 0$ (which is also the transition temperature). One dimension is also special because it is the only case for which the scaling function f which appears in (15) is known exactly (16). Another special feature of one dimension is that $m(t) = m(0)$.

Remark 4: the fact that the magnetization is conserved in one dimension (for Glauber dynamics) seems to be in apparent contradiction with the fact that domains grow. For example, for a finite system, the average magnetization is conserved (13) and it is clear that in the long time limit all the spins in the system are parallel (once the characteristic domain size is much larger than the system size). The solution to this paradox is that, in the long time limit, when all the spins in the system become parallel, they can be either all up or all down and the fact that the average magnetization is conserved implies that the spins become all up with probability $(1 + m(0))/2$ and all down with probability $(1 - m(0))/2$.

Remark 5: in dimension higher than 1, for Glauber dynamics, if the initial condition is weakly magnetized, the magnetization grows like

$$m(t) \sim [L(t)]^\lambda \, m(0)$$

where the exponent λ depends on dimension and is not known exactly except in $d = 1$ where $\lambda = 0$ (see 13) and in $d = 2$ where it has been conjectured [19] that $\lambda = 3/4$. The same exponent λ governs also the decay of the autocorrelation function [20, 22, 21, 23] in arbitrary dimension

$$\langle S_i(t) S_i(0) \rangle \sim [L(t)]^{-(d-\lambda)} \, .$$

A simple argument [24] can be used to explain why the same λ is present in both expressions: Consider a system of N Ising spins in dimension d and call $P(\{S_i(t)\} \,|\, \{S_i(0)\})$ the probability of finding the system in the spin configuration $\{S_i(t)\}$ at time t given that it was in configuration $\{S_i(0)\}$ at time 0. When the initial condition $\{S_i(0)\}$ is chosen completely at random, the correlation $\langle S_i(t)S_j(0)\rangle$ is given by

$$\langle S_i(t)S_j(0)\rangle = \frac{1}{2^N} \sum_{\{S(t)\}} \sum_{\{S(0)\}} S_i(t)S_j(0)\, P(\{S_i(t)\}\,|\,\{S_i(0)\})\,. \qquad (23)$$

On the other hand when one starts with a weakly magnetized initial condition, i.e. the initial configuration $\{S_i(0)\}$ is chosen with probability

$$Q(\{S_i(0)\}) = \prod_{i=1}^{N} \frac{1+m(0)S_i(0)}{2} \simeq \frac{1+m(0)\sum_j S_j(0)}{2^N}$$

with $m(0)$ infinitesimal, the magnetization $m(t)$ per spin at time t is to first order in powers of $m(0)$

$$m(t) = \sum_{\{S(t)\}} \sum_{\{S(0)\}} P(\{S_i(t)\}\,|\,\{S_i(0)\})\, Q(\{S_i(0)\}) \frac{\sum_j S_j(t)}{N}$$

$$\simeq m(0) \frac{\sum_i \sum_j \langle S_i(t)S_j(0)\rangle}{N}\,. \qquad (24)$$

If one assumes that due to some coarsening phenomenon the two-point function scales as

$$\langle S_i(0)S_j(t)\rangle \simeq L^{-(d-\lambda)} f(\frac{R_{ij}}{L})$$

where R_{ij} is the distance between sites i and j, one finds that

$$m(t) \simeq L^\lambda m(0) \int d^d R f(R)\,.$$

Thus the exponent λ governs both the magnetization and the autocorrelation.

Remark 6: in principle, one could try to generalize the above calculation (13,15,16) to obtain higher correlations. The knowledge of arbitrary correlations would, in particular, be useful to determine the distribution of domain sizes.

4 Zero temperature dynamics of the Potts model

The properties discussed so far for the Ising model can be generalized to the q-state Potts model. We shall see that for $q = \infty$, there are a few additional properties which can be calculated exactly.

Let us first discuss the dynamics of the q-state Potts model. Initially, each spin of the chain is given a color $S_i(0)$ at random ($1 \leq S_i(0) \leq q$). The Glauber dynamics at zero temperature can be implemented (as for the Ising model) by the following evolution rule: during each infinitesimal time interval dt,

$$S_i(t+dt) = \begin{cases} S_i(t) & \text{with probability } 1-dt \\ S_{i-1}(t) & \text{with probability } dt/2 \\ S_{i+1}(t) & \text{with probability } dt/2\,. \end{cases} \qquad (25)$$

This indicates that many dynamical properties of the one dimensional Potts model can be understood by random walks methods. Consider a random walker which, during each time interval dt, does not move with probability $1 - dt$ and jumps to its right or to its left with probability $dt/2$. The probability $P(y|x;t)$ of finding at time t this random walker at site y, given that it started at x is

$$P(y|x;t) = \sum_{n=0}^{\infty} e^{-t} \frac{t^n}{n!} \frac{n!}{\left(\frac{n+y-x}{2}\right)! \left(\frac{n+y+x}{2}\right)!}$$

$$= \frac{1}{2\pi} \int_0^{2\pi} d\theta \, \cos[(y-x)\theta] \, e^{-t(1-\cos\theta)} . \qquad (26)$$

An important property which will be useful later is the probability $Q(y|x;y',x';t)$ that two random walkers go from x to y and x' to y' (respectively) during time t without meeting. This can be obtained (for $(y'-y)(x'-x) > 0$) by the method of images

$$Q(y|x;y'|x';t) = P(y|x;t)P(y'|x';t) - P(y|x';t)P(y'|x;t) . \qquad (27)$$

From these properties, it is easy to extend to the q-state Potts model the results of the previous section. For example, the autocorrelation function is given by

$$\text{Prob}(S_i(t) = S_i(0)) = \frac{1 + (q-1)P(0|0;t)}{q}$$

$$= \frac{1}{q} + \frac{q-1}{q} \frac{1}{2\pi} \int_0^{2\pi} d\theta \, e^{-t(1-\cos\theta)} . \qquad (28)$$

Similarly, for an uncorrelated initial condition (i.e. such that $\text{Prob}(S_i(0) = S_j(0)) = 1/q$ for $i < j$), one has for the equal time two point function

$$\text{Prob}(S_i(t) = S_j(t)) = 1 - \frac{q-1}{q} \sum_{x<y} Q(x|i;y|j;t) .$$

In the long time limit, (26) becomes

$$P(y|x;t) = \frac{1}{\sqrt{2\pi t}} e^{-(y-x)^2/2t}$$

which leads to

$$\text{Prob}(S_i(t) = S_i(0)) = \frac{1}{q} + \frac{q-1}{q} \frac{1}{\sqrt{2\pi t}}$$

and for $i < j$

$$\text{Prob}(S_i(t) = S_j(t))$$

$$= 1 - \frac{q-1}{q} \frac{1}{2\pi t} \int_{-\infty}^{\infty} dx \int_x^{\infty} dy \left[e^{-\frac{(x-i)^2+(y-j)^2}{2t}} - e^{-\frac{(x-j)^2+(y-i)^2}{2t}} \right]$$

$$= 1 - \frac{q-1}{q} \frac{1}{\sqrt{\pi}} \int_0^{(j-i)/\sqrt{t}} e^{-u^2/4} du . \qquad (29)$$

This gives for the nearest neighbor correlation

$$\text{Prob}(S_i(t) = S_{i+1}(t)) = 1 - \frac{q-1}{q}\frac{1}{\sqrt{\pi t}} \; . \tag{30}$$

and as a consequence, the average number $\langle \nu_i(t) \rangle$ of flips of spin i is given by

$$\langle \nu_i(t) \rangle = \int_0^t dt' \, \frac{\text{Prob}(S_i(t') \neq S_{i+1}(t')) + \text{Prob}(S_i(t') \neq S_{i-1}(t'))}{2}$$

$$\simeq \frac{2(q-1)}{q} \sqrt{\frac{t}{\pi}} \; . \tag{31}$$

The case $q = \infty$ is simple enough to allow one to obtain two other interesting properties of the zero temperature dynamics. Let us first consider the case of no correlation in the initial condition (i.e. since $q = \infty$, initially all the spins are different). One can show that the average number $\langle \mu_i(t) \rangle$ of different values taken by spin i between time 0 and time t grows like

$$\langle \mu_i(t) \rangle \simeq \frac{2}{\pi} \log t \; . \tag{32}$$

This can also be understood in terms of random walks. If $p(x|x_0;t)$ is the probability for a random walk to go from x_0 to x during time t *without visiting the origin*, one can show (using once more the method of images) that

$$p(x|x_0;t) = \frac{1}{2\pi} \int_0^{2\pi} d\theta [\cos(x-x_0)\theta - \cos(x+x_0)\theta\,] \, e^{-t(1-\cos\theta)} \tag{33}$$

which for large t gives

$$p(x|x_0;t) \simeq \frac{1}{\sqrt{2\pi t}} \left[e^{-\frac{(x-x_0)^2}{2t}} - e^{-\frac{(x+x_0)^2}{2t}} \right] \; .$$

For $j > 0$, the probability $\psi_j(t)$ that site 0 takes at least once the color $S_j(0)$ between time 0 and time t evolves according to

$$\frac{d\psi_j}{dt} = \frac{1}{2} \sum_{z=2}^{\infty} p(1|j;t)p(z|j+1;t) - p(1|j+1;t)p(z|j;t) \; . \tag{34}$$

This can be understood by considering at time t the domain of spins which has the color $S_j(0)$: the two boundaries of this domain perform random walks and the formula (34) expresses the fact that one of the two domain walls hits the origin for the first time at time t (without having met the other domain wall between time 0 and time t). This gives for large t

$$\frac{d\psi_j}{dt} = \frac{1}{\pi t} \left[\frac{j}{t} e^{-\frac{j^2}{t}} + \left(\frac{j^2}{t^2} - \frac{1}{t} \right) e^{-\frac{j^2}{2t}} \int_0^j e^{-\frac{u^2}{2t}} du \right] \; .$$

and one obtains

$$\frac{d\langle \mu_i \rangle}{dt} = 2 \sum_{j>0} \frac{d\psi_j}{dt} \simeq \frac{2}{\pi t}$$

(the factor 2 being due to the contributions of the two sides of the origin). This implies that

$$\langle \mu_i(t) \rangle \simeq \frac{2}{\pi} \log t \ .$$

Remark: it is interesting to compare the average number of flips $\langle \nu_i(t) \rangle \simeq 2\sqrt{t/\pi}$ and the average number of different values $\langle \mu_i(t) \rangle \simeq (2/\pi) \log t$. The fact that $\langle \nu_i(t) \rangle \gg \langle \mu_i(t) \rangle$ means that when a spin flips, most of the times, it takes colors that it had already taken in the past.

Another property which can be obtained for the q-state Potts model when $q = \infty$ is the distribution of domain sizes. This again can be done using random walks. We start at $t = 0$ with an initial configuration where all spins have different colors, so that initially all domains have length 1. Due to the motion of its two domain walls, during each time dt, the length of a domain has a probability dt of decreasing by 1 and a probability dt of increasing by 1. Moreover, for a domain to survive up to time t, its length should not vanish between time 0 and time t. This gives for $\tilde{p}_n(t)$, the probability for a domain to be of length n at time t

$$\tilde{p}_n(t) = \frac{1}{2\pi} \int_0^{2\pi} d\theta \ [\cos(n-1)\theta \ - \ \cos(n+1)\theta] \ e^{-2t(1-\cos\theta)}$$

which for large t gives for $n \geq 1$

$$\tilde{p}_n(t) = \frac{n}{2\sqrt{\pi} \ t^{3/2}} \ e^{-\frac{n^2}{4t}} \ . \tag{35}$$

This distribution is not normalized because lots of domains disappear between time 0 and time t. In fact the number of domains per unit length decreases like $(\pi t)^{-1/2}$ in agreement with (30). If one normalizes (35), one finds [9] that the probability $p_n(t)$ for a domain to be of length n is

$$p_n(t) = \frac{\tilde{p}_n(t)}{\sum_{n' \geq 1} \tilde{p}_{n'}(t)} \simeq \frac{1}{t^{1/2}} \ F(n/\sqrt{t}) \tag{36}$$

with

$$F(x) = \frac{x}{2} e^{-\frac{x^2}{4}} \ . \tag{37}$$

Remark 1: for $q = \infty$, because all the domains have different colors, one can relate $p_n(t)$ and the correlation function $\text{Prob}(S_i(t) = S_j(t))$

$$\text{Prob}(S_i(t) = S_j(t)) = \frac{\sum_{n \geq |j-i|} [n - |j-i|] \ p_n(t)}{\sum_n n \ p_n(t)} \simeq \frac{1}{\sqrt{\pi}} \int_{|j-i|/\sqrt{t}}^{\infty} e^{-\frac{u^2}{4}} \ du \ .$$

When q is finite, there are contributions to $\text{Prob}(S_i(t) = S_j(t))$ coming from spins in different domains which make the relation between the distribution of domain sizes and the correlation functions more complicated.

Remark 2: the result (36-37) was obtained for an initial condition where all the domains have length 1. If one starts with a more general distribution of initial lengths, which decreases fast enough (i.e. $p_n(0)$ decays faster than $1/n^2$ for large n), the system reaches the same scaling limit characterized by the scaling function (37). On the other hand, if initially, the distribution $p_n(0)$ decays like a power law $n^{-1-\alpha}$ with $1 < \alpha < 2$, one can show [9] that the system reaches another scaling regime where the scaling function $F(x)$ is replaced by an F_α which depends on α

$$F_\alpha(x) = C^{te} \, e^{-x^2/4} \int_0^\infty du \, \frac{e^{-u^2} \sinh(ux)}{u^{1+\alpha}} \, .$$

5 Reaction diffusion models

Instead of working with the spin variables $S_i(t)$, one can also discuss the dynamics of the Ising or the Potts chain in terms of domain walls [40].

For the q-state Potts model, the domain walls evolve according to the following reaction diffusion model [25, 26, 27, 28, 29, 30, 31, 32, 33, 34, 35, 36, 37]. Particles diffuse on a one dimensional lattice (each particle performs a random walk which, during each time step dt moves to the right with probability $dt/2$ or to the left with probability $dt/2$). Whenever two particles occupy the same site, they instantaneously react according to

$$A + A \rightarrow \begin{cases} A & \text{with probability } \frac{q-2}{q-1} \\ \emptyset & \text{with probability } \frac{1}{q-1} \end{cases}$$

So when two particles meet, they either aggregate with probability $(q-2)/(q-1)$ to form a new particle A or they annihilate with probability $1/(q-1)$.

Two limiting cases are of interest: for $q = 2$ (Ising), particles always annihilate when they meet whereas for $q = \infty$, particles always aggregate when they meet.

Let us call τ_i the binary variable which indicates whether bond $i, i+1$ (domain walls are defined on bonds) is occupied by a particle ($\tau_i = 1$) or empty ($\tau_i = 0$). The dynamics defined above implies the following time evolution for τ_i:

$$\tau_i(t+dt) = \begin{cases} \tau_i & \text{with probability} \\ & 1 - 2dt \\ 1 & \text{with probability} \\ & \frac{dt}{2}\left[\tau_i(2 - \tau_{i+1} - \tau_{i-1}) + (\tau_{i+1} + \tau_{i-1})(1 - \tau_i)\right. \\ & \left. + \frac{q-2}{q-1}\tau_i(\tau_{i+1} + \tau_{i-1})\right] \\ 0 & \text{with probability} \\ & \frac{dt}{2}\left[2 + (1 - \tau_i)(2 - \tau_{i+1} - \tau_{i-1})\right. \\ & \left. + \frac{1}{q-1}\tau_i(\tau_{i+1} + \tau_{i-1})\right] \end{cases}$$

This leads to the following time evolution

$$\frac{d\langle\tau_i\rangle}{dt} = -\langle\tau_i\rangle + \frac{1}{2}(\langle\tau_{i-1}\rangle + \langle\tau_{i+1}\rangle) - \frac{q}{2(q-1)}(\langle\tau_i\tau_{i+1}\rangle + \langle\tau_i\tau_{i-1}\rangle) \, . \qquad (38)$$

The calculation of the time evolution of $\langle \tau_i \rangle$ looks much more difficult than in (3-6) because the evolution of the one point function requires the knowledge of the two point function $\langle \tau_i \tau_{i+1} \rangle$ which itself requires the knowledge of the three point function and so on

$$\frac{d\langle \tau_i \tau_{i+1} \rangle}{dt} = -2\langle \tau_i \tau_{i+1} \rangle + \frac{1}{2}(\langle \tau_i \tau_{i+2} \rangle$$

$$+ \langle \tau_{i-1} \tau_{i+1} \rangle) - \frac{q}{2(q-1)}(\langle \tau_i \tau_{i+1} \tau_{i+2} \rangle + \langle \tau_{i-1} \tau_i \tau_{i+1} \rangle) . \quad (39)$$

Thus, apparently, to know the one point function, one needs to calculate all the correlation functions at the same time and this usually is a very difficult task.

However, the link with the spin models makes the calculation of $\langle \tau_i \rangle$ very easy. The probability $\langle \tau_i \rangle$ that a domain wall is present between two consecutive sites i and $i+1$ is just $\text{Prob}(S_i(t) \neq S_{i+1}(t))$. Therefore, from (30), it is clear that in the long time limit, the density is given by

$$\langle \tau_i \rangle \simeq \frac{q-1}{q} \frac{1}{\sqrt{\pi t}} .$$

Remark 1: it is interesting to compare this result with the prediction of the mean field approximation obtained by replacing $\langle \tau_i \tau_k \rangle$ by $\langle \tau_i \rangle \langle \tau_k \rangle$ in (38)

$$\langle \tau_i \rangle \simeq \frac{q-1}{q} \frac{1}{t} . \quad (40)$$

Remark 2: this $1/t$ decay predicted by the mean field theory is expected to be valid for the reaction diffusion model when particles diffuse in dimension $d > 2$. For $d < 2$, one expects [25, 30, 31, 35] the density to decay like $t^{-d/2}$. This can be understood by the following argument: if the density of particles is ρ, the typical distance between particles is $a = \rho^{-1/d}$. Then since the particles diffuse, it takes at least a time $t \sim a^2$ for a particle to meet another one. So if one writes that $a \sim t^{1/2} \sim \rho^{-1/d}$, one finds that

$$\rho \sim t^{-d/2} .$$

6 The spins which never flip

For the zero temperature dynamics of the $1d$ Ising or Potts models, one can measure the density $r(q, t)$ of spins which have never flipped. One observes [14, 38] that for a random initial condition (where each spin is given at random one of the q colors), the density $r(q, t)$ decreases like a power law

$$r(q, t) \simeq t^{-\theta(q)} \quad (41)$$

where the exponent $\theta(q)$ varies with q. MonteCarlo simulations [14, 38, 39] give $\theta(2) \simeq 0.376$, $\theta(3) \simeq 0.53$, $\theta(5) \simeq 0.70$, $\theta(10) \simeq 0.82$.

In section 4, we considered for $q = \infty$, the number $\mu_i(t)$ of different values taken by a spin i between time 0 and time t. We saw (32) that $\langle \mu_i(t) \rangle \simeq \frac{2}{\pi} \log t$.

The only difference between the $q = \infty$ and the finite q cases is that the probability that two spins are different in the initial condition is 1 for $q = \infty$ and is $(q - 1)/q$ for $q < \infty$. Therefore if $\mu_i(t)$ is the different number of values taken by spin i for $q = \infty$, one has for $q < \infty$

$$r(q, t) = \langle q^{1-\mu_i(t)} \rangle \ . \tag{42}$$

The full distribution of $\mu_i(t)$ is not yet known. However [15] this expression together with (32) gives both the exponent $\theta(q)$ for q close to 1 and an upper bound for arbitrary q. For

$$q = 1 + \epsilon$$

one obtains from (42) and (32)

$$r(q, t) = 1 - \epsilon \langle (\mu_i(t) - 1) \rangle + O(\epsilon^2) \simeq 1 - \frac{2}{\pi} \epsilon \, \log t + O(\epsilon^2)$$

which implies that

$$\theta(q) = \frac{2}{\pi}(q - 1) + O[(q - 1)^2] \ . \tag{43}$$

On the other hand (42) implies, using Jensen's inequality, that

$$r(q, t) \geq q^{1-\langle \mu_i(t) \rangle}$$

which gives for all values of q, the following upper bound for $\theta(q)$

$$\theta(q) \leq \frac{2}{\pi} \log q \ .$$

Analytically, it is difficult to go further for general q. Only for $q = \infty$, one can use random walks ideas to solve exactly the problem. For $q = \infty$, all the spins are different in the initial condition. Thus for spin i not to flip between time 0 and time t, one needs that the two domain walls at its left and at its right do not cross i. This can be easily calculated from the probability (33) that a random walk does not visit the origin up to time t

$$r(\infty, t) = \left(\sum_{x \geq 1} p(x|1; t) \right)^2$$

which gives for large t

$$r(\infty, t) \simeq \frac{2}{\pi t}$$

so that

$$\theta(\infty) = 1 \ .$$

For cases where $\theta(q)$ is not known exactly, one needs to use approximate methods. One method which can be used and and has the advantage that it can be improved systematically is finite size scaling ([41, 42, 15]). One starts by computing exactly the properties of finite systems. Then one extrapolates these properties to obtain the behavior of the infinite system.

Here we consider a system of L spins on a ring. For a random initial condition, the dynamics always drives the system to a configuration where all spins are parallel. Let us call $\rho(q, L)$ the probability that between time 0 and time ∞, a given spin never flips. One can calculate exactly $\rho(q, L)$ by hand for very small L and one can go up to $L \simeq 14$ using a computer [15]. One finds

$$\rho(q, 1) = 1$$

$$\rho(q, 2) = \frac{q+1}{2q}$$

$$\rho(q, 3) = \frac{2q^2 + 4q + 1}{7q^2}$$

$$\rho(q, 4) = \frac{8q^3 + 23q^2 + 12q + 1}{44q^3}$$

$$\rho(q, 5) = \frac{62q^4 + 224q^3 + 176q^2 + 32q + 1}{495q^4}$$

$$\rho(q, 6) = \frac{912q^5 + 3864q^4 + 3983q^3 + 1135q^2 + 81q + 1}{9976q^5}$$

$$\rho(q, 7) = \frac{25086q^6 + 119732q^5 + 150210q^4 + 58176q^3 + 6756q^2 + 200q + 1}{360161q^6}$$

and so on.

To see how these results are obtained, let us consider the very simple case $L = 3$. There are three possibilities for the initial condition: (1) if initially the 3 spins are identical (probability $1/q^2$), spin 1 never moves; (2) if only one of the two other spins is identical to spin 1 (probability $2(q-1)/q^2$), spin 1 never moves with probability 4/7; (3) if both spins 2 and 3 are initially different from spin 1 (probability $(q - 1)^2/q^2$), spin 1 never moves with probability 2/7. Therefore

$$\rho(q, 3) = \frac{1}{q^2} + \frac{2(q - 1)}{q^2} \frac{4}{7} + \frac{(q - 1)^2}{q^2} \frac{2}{7} .$$

The difference between the finite and the infinite system shows up when the size $t^{1/2}$ of the growing domains becomes comparable to the system size L. For times $t \sim L^2$, the dynamics of the finite system stops because all the spins have become parallel. Therefore, one expects $\rho(q, L)$ to decay like

$$\rho(q, L) \sim t^{-\theta(q)} \sim L^{-2\theta(q)} .$$

This large L behavior allows one to estimate $\theta(q)$ from the knowledge of the $\rho(q, L)$ by the following expression

$$\theta(q) = -\frac{1}{2} \frac{\log[\rho(q, L + 1)/\rho(q, L)]}{\log[(L + 1)/L]} . \tag{44}$$

The results of this finite size scaling estimate are given for $q = 2$ in column 0 of table 1. The estimate (44) changes with L in a rather regular way. In columns 1,2,... the results of column 0 are extrapolated by using polynomials of degree 1,2,... in $1/L$.

	$q = 2$						
L	0	1	2	3	4	5	6
1	0.207519	0.313634	0.360881	0.371070	0.371517	0.373260	0.374975
2	0.260576	0.345132	0.368523	0.371428	0.372970	0.374730	0.375187
3	0.288761	0.356827	0.370266	0.372456	0.374227	0.375073	0.375037
4	0.305778	0.362203	0.371361	0.373408	0.374756	0.375049	0.374983
5	0.317063	0.365255	0.372264	0.374112	0.374918	0.375009	0.374989
6	0.325095	0.367258	0.372957	0.374470	0.374964	0.374998	0.375000
7	0.331118	0.368683	0.373461	0.374668	0.374980	0.374999	0.375004
8	0.335814	0.369744	0.373823	0.374781	0.374988	0.375001	
9	0.339584	0.370560	0.374084	0.374850	0.374993		
10	0.342681	0.371201	0.374276	0.374894			
11	0.345274	0.371713	0.374418				
12	0.347477	0.372130					
13	0.349374						

Table 1. Estimates of the exponent $\theta(q)$ for $q = 2$. The column 0 is obtained from (41). Columns 1-6 are polynomial extrapolations of degree 1-6 of the data of column 0.

We see that for $q = 2$, the extrapolation procedure leads to the following estimate for $\theta(q)$

$$\theta(2) = 0.3750 \pm 0.0001 . \tag{45}$$

Using the same procedure for $q = \infty$, one recovers (see table 2) the exactly known result $\theta(\infty) = 1$ quite accurately. Of course, as q appears just as a parameter, one can repeat the finite size scaling calculation for other values of q. Some results are given in table 3.

The mean field theory is another possible approximation which can be done for this problem. We have already obtained (38) the time evolution of the probability $\langle \tau_i \rangle$ of having a domain wall between site i and site $i + 1$

$$\frac{d\langle \tau_i \rangle}{dt} = -\langle \tau_i \rangle + \frac{1}{2}(\langle \tau_{i-1} \rangle + \langle \tau_{i+1} \rangle) - \frac{q}{2(q-1)}(\langle \tau_i \tau_{i+1} \rangle + \langle \tau_i \tau_{i-1} \rangle) . \tag{46}$$

If one introduces the variable η_i on site i to distinguish between the fact that spin i has never flipped up to time t ($\eta_i = 1$) and the fact that it has flipped at least once ($\eta_i = 0$), the exact time evolution of $\langle \eta_i \rangle$ is given by

$$\frac{d\langle \eta_i \rangle}{dt} = -\frac{1}{2} (\langle \eta_i \tau_i \rangle + \langle \eta_i \tau_{i-1} \rangle) . \tag{47}$$

L	0	1	2	3	4	5	6
				$q=\infty$			
1	0.500000	0.880182	1.024671	0.979200	0.946464	0.988116	1.016309
2	0.690091	0.976508	0.990568	0.953012	0.981174	1.012281	1.004275
3	0.785564	0.983538	0.968034	0.971786	1.003393	1.006277	0.996287
4	0.835057	0.977337	0.969910	0.989848	1.005196	0.999617	0.997456
5	0.863513	0.974861	0.978455	0.997522	1.002096	0.998321	0.999578
6	0.882071	0.975888	0.985605	0.999555	1.000209	0.999007	1.000297
7	0.895474	0.978317	0.990255	0.999816	0.999662	0.999652	1.000272
8	0.905829	0.980970	0.993123	0.999760	0.999658	0.999938	
9	0.914178	0.983401	0.994933	0.999726	0.999766		
10	0.921100	0.985498	0.996132	0.999738			
11	0.926955	0.987270	0.996964				
12	0.931981	0.988761					
13	0.936349						

Table 2. Same results as in table 1 for $q = \infty$.

q	.5	1.5	2	3	5	10	100	∞
$\theta(q)$	-.513	.2350	.3750	.5379	.6928	.8310	.9815	1.0000
Error	.001	.0001	.0001	.0002	.0003	.0005	.0010	.0005
Monte Carlo			.376	.53	.70	.82		1.0

Table 3. Estimates of $\theta(q)$ obtained using finite size scaling.

Both (46) and (47) are exact. Unfortunately they both involve higher order correlations and this makes the problem very difficult.

The mean field approximation consists in neglecting correlations (i.e. in replacing $\langle \tau_i \tau_k \rangle$ by $\langle \tau_i \rangle \langle \tau_k \rangle$ and $\langle \eta_i \tau_k \rangle$ by $\langle \eta_i \rangle \langle \tau_k \rangle$ in (46) and (47)). Within this approximation, one finds (40) for large t

$$\langle \tau_i \rangle \simeq \frac{q-1}{q} \frac{1}{t}$$

and

$$\frac{d\langle \eta_i \rangle}{dt} = -\frac{q-1}{q} \frac{1}{t} \langle \eta_i \rangle$$

which gives

$$\langle \eta_i(t) \rangle \simeq t^{(q-1)/q} .$$

So we see that the mean field approximation predicts an exponent

$$\theta_{\text{mean field}}(q) = \frac{q-1}{q} \tag{48}$$

which varies with q.

Remark 1: The mean field approximation gives an exponent (48) which is certainly not correct. For $q = 1 + \epsilon$, it disagrees with (43) and for other values of q, it also disagrees with the finite size scaling predictions of table 3 (for example, the finite size scaling calculation gives $\theta(2) \simeq 3/8$ instead of $\theta_{mean\ field} = 1/2$). However, it is interesting to note that for this problem, the mean field theory is able to predict an exponent which varies continuously with q.

Remark 2: The exponent θ can also be measured in higher dimension [14, 38]. For the zero temperature Glauber dynamics of the Ising model, $\theta \simeq 0.220$ in $d = 2$ and $\theta \simeq 0.166$ in $d = 3$.

7 Conclusion

In this lecture we have seen that the one dimensional Ising or Potts model at zero temperature exhibit rather interesting dynamical properties: coarsening (15), aging (22), non trivial exponents for the number of spins which never flip (41). Most properties can be formulated in terms of random walks or of reaction diffusion models. For example using random walks, it is rather easy to show that starting from a random initial condition, the system reaches, in the long time limit, a scaling regime characterised by a single length which grows as $t^{1/2}$. Other properties like the calculation of the exponent $\theta(q)$ which governs the fraction of spins which have never flipped (41) are more difficult to understand. However, a recent approach developped in collaboration with Vincent Hakim and Vincent Pasquier [43] allowes one to calculate analytically the $\rho(q, L)$ of section 6 for arbitrary large L. This should lead to an exact expression of the exponent $\theta(q)$.

There are other one dimensional models which exhibit coarsening phenomena. One of them is the one dimensional Ginzburg-Landau equation [44]. A configuration of the system at time t is represented by a function $\Phi(x, t)$ and it evolves deterministically according to

$$\frac{\partial \Phi}{\partial t} = \frac{\partial^2 \Phi}{\partial x^2} + \Phi - \Phi^3 . \tag{49}$$

Regions where Φ is positive correspond to the $+$ phase whereas Φ is negative in the $-$ phase. A major difference with the zero temperature dynamics of the Ising chain discussed in section 3 is that here the dynamics is deterministic (49) and there is only randomness in the initial condition. If one starts with a random initial condition where $-1 \leq \Phi(x, 0) \leq 1$, one observes a coarsening phenomena. In the long time limit, $\Phi(x) = 1$ or -1 almost everywhere except at domain walls. As the domains grow, the typical size of the domains becomes much larger than the width of domain walls. In this limit, the dynamics becomes very simple: the closest pair of walls move together and annihilate while the other walls hardly move at all and the system coarsens by eliminating the smallest domains [44, 45].

In the scaling limit, one can calculate exactly the distribution of domain sizes [44, 46, 47]. One can also show [24] that the exponent λ (which governs the autocorrelation function and the magnetization as discussed in the last remark of section 3) is exactly given by the zero of the following transcendental equation

$$\int_0^\infty q^{\lambda-2} e^{-q} \left[(1 - q - e^{-q}) e^{r(q)} + q^2 (1 - \lambda) e^{-r(q)} \right] dq = 0 \qquad (50)$$

where

$$r(p) = \int_p^\infty \frac{e^{-q}}{q} dq + \log p , \qquad (51)$$

leading to

$$\lambda \simeq .3993835..$$

One can also show [47] that the fraction of the line which has never flipped decays like $L^{-.1750758..}$ where L is the characteristic size of the domains. It is remarkable the solution of this simple model of coarsening leads to such nontrivial exponents.

This lecture has been inspired by very pleasnt and stimulating interactions with A. J. Bray, D. Dhar, M. Droz, H. Flyvbjerg, C. Godrèche, V. Hakim, V. Pasquier and I. Yekutieli.

While it was in press, the approach developped in [43] led to an exact analytic expression for $\theta(q)$

$$\theta(q) = -\frac{1}{8} + \frac{2}{\pi^2} \left[\cos^{-1} \left(\frac{2 - q}{\sqrt{2} \, q} \right) \right]^2$$

and this expression is in excellent agreement with the numerical estimates of section 6.

References

1. A. J. Bray, Advances in Physics **44**, 357 (1994)
2. P. S. Sahni, D.J. Srolovitz, G.S. Grest, M. P. Anderson and S.A. Safran, Phys. Rev. B **28**, 2705 (1983).
3. M. P. Anderson, D.J. Srolovitz, G.S. Grest and P. S. Sahni, Acta Metall. **32**, 783 (1984).
4. C. Sire and S. N. Majumdar, Phys. Rev. Lett. **74**, 4321 (1995).
5. H. Flyvbjerg, Phys. Rev. E **47**, 4037 (1993).
6. J. Stavans, E. Domany and D. Mukamel, Europhys. Lett. **15**, 479 (1991).
7. D. Fritter, C.M. Knobler, D. Roux and D. Beysens J. Stat. Phys. **52**, 1447 (1988).
8. F. Family and P. Meakin, Phys. Rev. Lett. **61**, 428 (1988); Phys. Rev. A **40**, 3836 (1989).
9. B. Derrida, C. Godrèche and I. Yekutieli, Phys. Rev. A **44**, 6241 (1991).
10. Z. Rácz, Phys. Rev. Lett. **55**, 1707 (1985).
11. A. J. Bray, J. Phys. A **23**, L67 (1990).

12. J. G. Amar and F. Family, Phys. Rev. A **41**, 3258 (1990).
13. V. Privman, J. Stat. Phys. **69**, 629 (1992).
14. B. Derrida, A.J. Bray and C. Godrèche, J. Phys. A **27**, L357 (1994).
15. B. Derrida, J. Phys. A **28**, 1481 (1995).
16. R. J. Glauber, J. Mat. Phys. **4**, 294 (1963).
17. M. Alba, M. Ocio and J. Hammann, Europhys. Lett. **2**, 45 (1986).
18. J. P. Bouchaud, J. Physique I 2, 1705 (1992) and references therein.
19. D. S. Fisher and D. A. Huse, Phys. Rev. B **38**, 373 (1988).
20. H. Furukawa, J. Phys. Soc. Jpn. **58**, 216 (1989); Phys. Rev. B **40**, 2341 (1989).
21. S. N. Majumdar and D. A. Huse, preprint 1994.
22. S. N. Majumdar and D. A. Huse and B. D. Lubachevsky, Phys. Rev. Lett. **73**, 182 (1994).
23. A. J. Bray and J. G. Kissner, J. Phys. A **25**, 31 (1992).
24. A. J. Bray, B. Derrida, Phys. Rev. E **51**, 1633(1995).
25. K. Kang and S. Redner, Phys. Rev. Lett. **52**, 955 (1984); Phys. Rev. A **32**, 435 (1985).
26. L. Peliti, J. Physique **46**, 1469 (1985).
27. C.R. Doering and D. ben-Avraham, Phys. Rev. A **38**, 3035 (1988); Phys. Rev. Lett. **62**, 2563 (1989).
28. D. ben-Avraham, M.A. Burschka and C. R. Doering, J. Stat. Phys. **60**, 695 (1990).
29. F. Leyvraz and S. Redner, Phys. Rev. Lett. **66**, 2168 (1991); Phys. Rev. A **46**, 3132 (1992).
30. M. Bramson and J. L. Lebowitz, J. Stat. Phys. **62**, 297 (1991).
31. S. Cornell, M. Droz and B. Chopard, Physica A **188**, 322 (1992).
32. R. B. Stinchcombe, M. D. Grynberg and M. Barma, Phys. Rev. E **47**, 4018 (1993).
33. F.C. Alcaraz, M. Droz, M. Henkel and V. Rittenberg, Ann. Phys. **230**,250 (1994).
34. I. Peschel, V. Rittenberg and U. Schultze, Nucl. Phys. B **430** [FS], 633 (1994).
35. B. P. Lee, J. Phys. A **27**,2633 (1994).
36. T. J. Newman, J. Phys. A **28**, L183 (1995).
37. G. M. Schütz, J. Stat. Phys. **79**, 243 (1995).
38. D. Stauffer J. Phys. A **27**, 5029 (1994).
39. P.L. Krapivsky, E. ben-Naim and S. Redner, Phys. Rev. E **50**, 2474 (1994).
40. J. L. Cardy, J. Phys. A **28**, L19 (1995).
41. J. L. Cardy, "Finite Size Scaling", North Holland (1988).
42. V. Privman, "Finite Size Scaling and Numerical Simulation of Statistical Mechanics", World Scientific (1990).
43. B. Derrida, V. Hakim and V. Pasquier, Phys. Rev. Lett. (1995) in press.
44. T. Nagai and K. Kawasaki, Physica A **134**, 483 (1986).
45. J. Carr and R. Pego, Proc. Roy. Soc. London A **436**, 569 (1992).
46. A. D. Rutenberg and A. J. Bray, Phys. Rev. E **50**, 1900 (1994).
47. A. J. Bray, B. Derrida and C. Godrèche, Europhys. Lett. **27**, 175 (1994).

Part VI
Cosmology as a Problem in Critical Phenomena

Cosmology as a problem in critical phenomena

Lee Smolin

Center for Gravitational Physics and Geometry, Department of Physics
Pennsylvania State University
University Park, Pa16802-6360, USA
smolin@phys.psu.edu   * permanent address
                    and
School of Natural Sciences, Institute for Advanced Study
Princeton, New Jersey, 08540, USA

Abstract. Several problems in cosmology and astrophysics are described in which critical phenomena of various types may play a role. These include the organization of the disks of spiral galaxies, various aspects of the problem of structure formation in cosmology, the problem of the selection of initial conditions and parameters in particle physics and cosmology and the problem of recovering the classical limit from non-perturbative formulations of quantum gravity.

A measure of complexity which is suggested by these applications, but which may also have application to other problems, is described.

1 Introduction

Until recently, most attempts to construct theories of physics and cosmology have begun with the point of view that the universe is, in its fundamentals, not very complicated. Unfortunately, it seems that the world often frustrates our desire to understand it simply: Ω [1] must originally have been very finely tuned originally to be close to one now, but the best evidence is that it is now in fact measurably less than one [1, 2]. The neutrinos are very light compared to every other mass scale, but there is evidence that they are not exactly massless [3], while the proton and neutron have almost, but not exactly, the same mass. Similarly, the cosmic microwave radiation that gives us a snapshot of conditions when the universe was a thousand times smaller is a black body to incredible precision, and is isotropic to a precision of around a part in 10^5 [4]; while at the present time, surveys of the actual distribution of matter show a world which has structure up to the largest scales that have been accurately mapped [5].

The last thirty years have indeed been an incredibly surprising and exciting time in cosmology and theoretical physics. At the risk of oversimplifying, it seems that our attempts to model the universe on both cosmological and microscopic scales are leading to the conclusion that the universe is much more intricately structured than was imagined in the nineteen-sixties. There is always risk in

---

[1] The average density of the universe is units of the amount needed for the characteristic time scale of the cosmic expansion to be infinite.

generalizations, but if one looks for them, it might be said that three themes have emerged during this period that characterize the direction in which we seem to be headed in both cosmology and elementary particle theory.

Complexity On many different scales, we are discovering that the universe is much more complex than we might have expected based on earlier theoretical ideas. At the largest scales, the distribution of galaxies in space shows structure that was largely unexpected [5], whose origins are still not satisfactorily understood. Finally, as I will describe, the galaxies themselves seem to be much more complex than might have been expected.

At the smallest scales, with the discovery of the charmed, bottom and top quarks, and the tau leptons, the number of fundamental particles has just about doubled since 1970. But we still have no understanding of the spectrum of fundamental fermions, nor do we have a theory that explains the eighteen or so parameters in the standard model. The spectra of masses and mixing angles shows a complexity that is rather puzzling, with up and down quarks quite light on the hadron scale, while the others are spread over a range of masses up to almost $200Gev$. The pattern of mixing angles is also rather complex, and we have to understand funny things like why parity seems to be so well respected by all of the interactions but one, which breaks it maximally, or why CP is broken, but just a bit. Whatever pattern of symmetry breaking is behind all of this, it is unlikely to be simple. The models of grand unification that are now being considered are correspondingly rather more complicated than the original $SU(5)$ theory, that had to be discarded because proton decay, if it takes place, is rarer than that theory naturally predicted. Unification has turned out to be a harder problem than perhaps it seemed in 1975, partly because the properties of the elementary particles and forces are themselves so diverse.

Furthermore, it seems that the world on every scale larger than the nucleus is much more complex, given the actual values of the masses, coupling constants and mixing angles, than would be the case were they to take most other values [7, 8, 9, 10]. For example, the fact that there are many different stable bound states of protons and neutrons seems due to several coincidences in the values of these parameters. It may even be said that the complexity of the world on astronomical scales is to some extent a consequence of the complexity of the spectrum of elementary particles and forces. For most other values of these parameters the chemistry, atomic and nuclear physics and astronomy of the world would be much simpler.

Hierarchies and approximate scale invariance In both cosmology and elementary particle physics, the basic units of structure are spread over many orders of magnitude in scale, and notions of approximate scale invariance play an important role. Perhaps the most basic unsolved problem in elementary particle physics is the hierarchy problem, which is to explain why there are such large ratios among the fundamental scales in physics. In the fundamental Planck units, the mass of the proton is 10^{-19}, the electron is three orders of magnitude smaller, and the cosmological constant is at most 10^{-60}.

Furthermore, the fact that the astronomical world shows structure on such a wide range of scales is a direct consequence of this hierarchy in fundamental physics. The typical mass of a star is given by

$$M_{Chandra} = m_{proton} \left(\frac{m_{Planck}}{m_{proton}} \right)^3 ,$$
(52)

while its lifetime is given by

$$t_{star} \approx \alpha \epsilon \frac{m_{proton}}{m_{electron}} \left(\frac{M_{Chandra}}{M} \right)^2 \left(\frac{m_{Planck}}{m_{proton}} \right)^3 t_{Planck}$$

$$\approx \left(\frac{M_{Chandra}}{M} \right)^2 10^{10} \ years$$
(53)

where $\epsilon \approx .007$ is the fusion efficiency.

It has also been estimated that the typical mass of a galaxy must be [8, 9],

$$M_{galactic} = m_{proton} \alpha^5 \sqrt{\frac{m_{proton}}{m_{electron}}} \left(\frac{m_{Planck}}{m_{proton}} \right)^3$$
(54)

Thus, the hierarchical structure we see in astronomy, with stars organized into much larger galaxies, which seem in turn to be collected in still larger structures, is actually a consequence of the hierarchy among the scales in fundamental physics.

There is also evidence that approximate scale invariance characterizes the distribution of galaxies in space, at least over a certain range of scales [43]. In addition, the most successful hypotheses for the initial fluctuations in mass density that ultimately lead to the formation of the galaxies and the large scale structure is that their distribution is scale invariant [43].

Going back to the small scale structure, because the fundamental length, the Planck length, is so small compared to the scales of strong interaction physics, the ground state of elementary particle physics is characterized by an approximate scale invariance at all scales larger than l_{Planck}. This has led to important conceptual tools in elementary particle theory, such as the renormalization group and the analogy between a quantum field theory and a statistical mechanical system at a second order critical point.

Evolution The most important way in which twentieth century cosmology differs from the Newtonian and Aristotelian cosmologies is that it is based on the understanding that the universe has evolved dramatically over time. Whatever happens concerning the details of the very early universe and the problems of structure formation, the successes of the big bang model, together with the failure of the steady state theory, leave us with a universe whose present state must be understood to be the result of physical processes which occurred at earlier times, when it was very much different. Thus, cosmology has become an historical science, in which a detailed story of what happened at earlier times has replaced the philosophical and *a priori* speculations that characterized most previous attempts at cosmology.

The notion of evolution has not so far played a correspondingly central role in elementary particle physics. This may, on reflection, seem unnatural, given the close relationship that is developing between particle physics and cosmology. Certainly, one must wonder what the traditional notion that the laws of physics represent timeless truths means in a universe whose origin we can literally almost see.

In the body of these notes I will elaborate on some implications of these three themes. Before beginning, however, some general comments are in order.

1.1 Why critical phenomena may be important for particle physics and cosmology

Since the 1970's there has been a mutually fruitful interaction between statistical mechanics and elementary particle physics, based largely on the formal analogy between second order phase transitions and the problem of renormalization in quantum field theory. At the root of this, however, is a deep problem for elementary particle physics, for this analogy is based on the fact that there is a fine tuning problem in quantum field theory. The parameters that specify the dynamics must be precisely tuned as a function of the cutoff scale, if there are to be interacting particles on scales much larger than the cutoff. It has helped a great deal to understand that this problem is analogous to the problem of tuning a statistical system to a critical point to describe a second order phase transition, but it does not solve the basic problem of why a fine tuning is needed in quantum field theory.

We may note that as long as renormalization is thought of as a mathematical process in which the cutoff energy scale is taken to infinity, then the fine tuning problem I have been speaking about is formal, as it concerns a technique used to construct the theory, and does not describe any phenomena in nature. But one thing all at least partially successful approaches to quantum gravity agree about is that the Planck scale does function as an effective short distance cutoff [6]. For apparently different reasons this is the case both in string theory and in the nonperturbative approaches to diffeomorphism invariant quantum field theories[2]. Once there is a physical cutoff the analogy between statistical mechanics and Euclidean quantum field theory becomes perfect and the fine tuning problem becomes a physical problem. It then becomes a problem of physics, and of critical phenomena in particular, to understand why our world has light particles in it.

More recently, a second domain of critical phenomena has come to light in statistical physics, in which no fine tuning is necessary [11, 12]. These are self-organized critical systems, which are non-equilibrium systems that spontaneously organized themselves in configurations characterized by approximate scale invariance over a wide range of scales, without the need for any precise tuning of parameters. It is then natural to ask whether such mechanisms, or some

---

[2] Of course, many people have hypothesized the existence of such a fundamental scale, what is significant is that it comes out of these two approaches to combining quantum theory and relativity without being put in by hand.

general mechanism of self-organization, might also play a role in elementary particle physics, to explain the fine tunings, and the existence of large hierarchies, that we now must impose.

Critical phenomena associated with phase transitions have also played a role in early universe cosmology. The two best studied ideas to explain the ultimate origin of the large scale structure, inflation and cosmic strings, involve phase transitions as the universe expands and cools. Both of these can lead to scale invariant distributions of initial fluctuations of the type that seem necessary to explain the current data about the large scale structure. However, in spite of these successes, there are indications that the models which have been studied so far may not in the end account completely for the large scale structure that is seen. The most important reason for this is that, as I mentioned in my opening, the evidence is more and more pointing to an Ω less than one. It is then natural to ask whether the more recently studied self-organized critical phenomena might play some role in the early history of the universe, and whether this might provide an alternative framework for understanding structure formation, and the origin of approximate scale invariance, in the large scale structure of the universe.

There have already been several proposals about how the statistical mechanics of self-organized systems may play a role in astrophysics. There are conjectures that the spectra of radiation coming from accretion disks around neutron stars or black holes might arise from self-organized critical systems [13]. In addition, there are suggestions that spiral galaxies may be described as stable non-equilibrium systems which are self-organized by the action of certain feedback processes involving star formation. These examples suggest that there may be fruitful scope for applying the physics of non-equilibrium and self-organized systems to problems in astrophysics and cosmology.

But perhaps more generally, I would like to propose that there must be a role for the physics of self-organized systems in cosmology and particle physics, simply because of the fact that it is highly non-trivial that the universe is as organized as it is. If it is the case that for most values of the parameters of particle physics and cosmology, and most choices of initial conditions, the universe would be much less varied and organized than it is presently, then there must be some reason for this. Given the incomplete success of other hypotheses, it perhaps is not inappropriate to begin to look for new ideas about the choices of parameters and initial conditions according to which the fact that the world is so organized may turn out to be essential rather than accidental.

But, if we seek a scientific explanation for this circumstance, then we have no recourse except in the physics of self-organized systems. The anthropic principle won't help us, for it assumes what we want to explain, which is that the universe is sufficiently intricately organized and out of equilibrium that life may exist. There is nothing outside the universe, by definition, so any processes that have acted in the past to organize it must be processes of self-organization.

Furthermore, due to the advances in the theory of self-organized systems due to Per Bak and his collaborators, we now know that self-organized systems are often critical systems, with structure spread out in space and time over every available scale. The fact that the distribution of matter in our universe is approximately scale invariant over many orders of magnitude suggests that it may be fruitful to seek to apply some of the ideas and strategies developed in the study of self-organized systems to unsolved problems in cosmology and astrophysics.

These notes are then meant as an introduction to several different problems in astrophysics and cosmology in which critical phenomena might plausibly play an important role. I begin in section 2 with the problem of the organization of spiral galaxies and then in the next two sections describe the open problems in our understanding of the large scale organization of the observed universe. All the facts presented in these three sections will be familiar to astronomers, even if the point of view may be nonstandard[3]. The last three then sections concern ways in which critical phenomena or mechanisms of self-organization may play a role in elementary particle theory, quantum gravity and general relativity.

2 Spiral galaxies as self-organized systems

A good place to start the discussion is with the disks of spiral galaxies, as this is one astrophysical domain in which it is clear that non-equilibrium processes are responsible for the formation and maintenance of structure. For this reason, it is also the one domain of astronomy that has been attacked in a serious way by physicists using the tools of modern statistical physics such as percolation theory and cellular automata. In a series of very interesting papers, Seiden, Schulman and Gerola constructed a theory of spiral structure based on an understanding of star formation as a certain kind of percolation process that spreads through the disk of the galaxy [14, 15]. To introduce the basic ideas of their theory I need first to review some of the basic facts about stars and galaxies.

A spiral galaxy, such as our own, consists of a number of components which are characterized by a surprising variety of structures and processes. The galaxy is surrounded by a spherical halo consisting primarily of old stars, as well as some unknown form of non-luminous matter. This dark matter seems to provide about $80 - 90\%$ of the mass, but does not otherwise participate in the energetics of the galaxy. For the moment we may leave to one side the very interesting question of its composition and origin.

Embedded in this halo is a disk consisting of gas, dust and stars of all ages. It is here that the dynamical processes that distinguish a spiral galaxy take place, and this will be the primary focus of this discussion. In the center of the disk is a bulge, which, like the halo, consists primarily of old stars. In the galaxies we will be concerned with here, the disk is much larger than the bulge.

The disk of a spiral galaxy seems to be a system which exists in a steady state, far from equilibrium, which is maintained by processes which cycle matter and energy among its various components [16, 17, 18, 19] [20, 21]. The evidence

---

[3] I apologize that, as these notes are intended as an introduction to these areas, and not as a comprehensive survey, no attempt has been made to provide a complete set of references.

is that the rates of these flows are approximately constant, averaged over the whole disk of the galaxy. Not surprisingly, some astronomers have proposed that there are feedback processes that govern the rates of flows of these cycles [16, 21]. To understand them we first must be familiar with the basic components and processes that make it up.

Stars come in a range of masses, from about $1/10$ to $100 M_{solar}$, where $M_{solar} = 2 \times 10^{33} grams$ is the mass of our sun. It is important to know that the luminosity of a star increases like the cube of its mass, so that the more massive stars dominate the energetics of the galactic disks. However, the lifetime decreases drastically, scaling like $mass^{-2}$. Because of these two facts, the stars of different masses play very different roles in the system of a galaxy. One basic fact is that the brighter and more massive stars radiate predominantly in the ultraviolet, so that they appear blue, while the less massive ones radiate primarily in the red.

Our understanding of the processes by which stars are formed is growing very rapidly at the present time [22]. What is clear is that, at least in spiral galaxies like our own, stars form in certain clouds of gas and dust called giant molecular clouds. We will speak about these shortly. A second very important fact is that the stars are created with a distribution of masses which is approximately a power law. This distribution is called *the initial mass function*, or IMF [23, 24]. Many more lower mass stars are formed originally; there is an empirical power law, due to Salpeter, that the number of stars born with mass between m and $m + dm$ scales like $m^{-\gamma}$ with γ a power between 2 and 3. There is evidence for a cut off on the low end, so that stars smaller than about $1/20$ of a solar mass are rare. There is also controversial evidence that the powers are different for low mass and high mass stars, which would suggest that they are formed in different processes [25].

Astronomers have looked for evidence that this initial mass function has varied over the lifetime of the galaxy or differs among galaxies; none has so far been found [24].

As a result of this, together with the fact that the more massive stars live for short times, the population of stars is dominated by the low mass stars. But, where they are found, the energetics is dominated by the massive stars.

The lower mass stars have lifetimes comparable to the present life of the universe (10^{10} years). When they burn out they end up quietly as a white dwarf. However, those stars more massive than about $8 M_{solar}$ end as supernovae, by which they expel all but about $1 - 3$ M_{solar}'s of their mass. The supernovaE also contribute a great deal of energy to the galaxy. These massive stars live for much shorter periods, with the time between formation and supernova typically on the order of 10^7 years. As this is much less than the rotation time of the galaxy (which is of order 10^8 years), this means that massive stars are found only in or near regions where star formation is taking place.

The spiral patterns one sees looking at a galactic disk are primarily caused by the very bright, massive stars. As such, these patterns trace the process of star formation. The disks apparently manifest spiral structure for the life of a galaxy, which is at least 10^{10} years. This is, at least in some galaxies, connected to the fact that the star formation rate is constant[4].

There are other processes besides supernova by which stars return matter to the interstellar medium of gas and dust out of which they are born. Massive stars evaporate a significant portion of their mass, this is the primary origin of the dust.

The dust and the gas together make up a clumpy medium which collects at the midplane of the disk. As a layer of gas, there is growing evidence that the disk extends far beyond the disk of stars. In the inner region containing the stars, the interstellar medium exists in several distinct phases, with greatly varying temperatures and pressures. To understand the role of the medium we need to describe these different phases [18, 20].

Most of the volume of the medium is taking up by a very rarified phase of hot ionized gas, with temperatures of greater than $10^5 K$. These are regions that have been evacuated and ionized by the passage of shock waves from supernovae. Next, going down in temperature, is a phase of warm gas, with temperatures on the order of 10^3 degrees K and densities on the order of one atom per cm^3. The gas is primarily atomic hydrogen.

Embedded in this warm gas are denser clouds, which apparently are continually condensing out of it. These clouds range from $10 - 100$ degree kelvin, with densities that range inversely from one up to 10^4 atoms per cm^3. In the denser and colder clouds there is a lot of dust, which apparently plays a role shielding the cloud from the ultraviolet light that would heat it. Because of this shielding, the gas in the clouds is molecular. Not only is the hydrogen in molecular form, but an array of organic elements are found there, including not only CO and NH_3 but alcohols and larger organic molecules. Because of this these are called the giant molecular clouds. They have masses on the order of 10^6 solar masses, and diameters of a few light years.

The distribution of matter within these giant molecular clouds is very irregular. There are suggestions that they have a filamentary structure; there are also suggestions that the distribution of densities in them is scale invariant up to large scales [26].

The giant molecular clouds play a key role in the galaxy because it is in them that the stars form.

The most important thing to understand about the star formation process in the giant molecular clouds is that it is rather inefficient [22]. This seems to be true for three reasons. First, a star begins to form when a cold and dense core of a cloud collapses. At some point its center is dense enough to ignite. This fuels a wind, which blows out from the star, or from an accretion disk surrounding it, which blows away the matter around the star, cutting off the accretion of matter onto the star. It is likely that this feedback process is responsible for the fact

---

[4] The spirals with constant star formation rate are type Sc, which have the largest ratio of the size of the disk relative to the bulge. In galaxies with much larger bulges the total star formation rate is now less than it was in the past. This suggests that a constant star formation rate is a property associated with the disk.

that the typical mass of a star is in fact just right for nuclear burning.

Second, when massive stars, form in a cloud, they heat it which after sufficient energy has been deposited in the cloud, apparently curtails further star formation[5]. Thus, there is a kind of a feedback effect which limits the efficiency of conversion of the giant molecular clouds to stars. Indeed, the very massive stars radiate in the ultraviolet, which ionizes the gas around them. These hot, ionized regions are found surrounding sites of recent massive star formation.

Third, while the clouds are dense and cold enough to collapse gravitationally, it seems that they are supported against collapse by some combination of turbulence and magnetic fields. This means that the rate of star formation can be greatly accelerated if the cloud is subject to an external perturbation such as a shock wave. Indeed, while low mass stars may spontaneously condense out of the giant molecular clouds, it is widely believed that the formation of massive stars would be much rarer in the absence of these external perturbations.

The main source for these external perturbations is believed to be other recently formed massive stars [19]. Primarily through supernovae, but also possibly through their ultraviolet radiation, very massive stars form shock waves in the interstellar medium. While these may destroy the giant molecular clouds in which they form, the result seems to be the catalysis of massive star formation in nearby giant molecular clouds.

This gives rise to a phenomena which is called self-propagating star formation [14]. As long as there is a continual supply of giant molecular clouds, the formation of massive stars can spread through the disk through a process in which the supernova of a massive star in one cloud catalyzes the formation of new massive stars in nearby clouds. The time scale for this process is the lifetime of a massive star, which is at most 10^7 years.

There are several independent pieces of evidence that the rate of star formation in the disk is governed by feedback processes occurring at several scales. The first is simply the fact that the interstellar medium maintains a configuration consisting of a number of different phases with approximately constant proportions of mass and volume. This is a dynamical stability, as the presence of the different phases means that the medium is out of equilibrium. Moreover, the giant molecular clouds must be condensing out of the warm gas as a steady rate, as they are being continually destroyed through the process of star formation. Further, the evidence shows that the star formation rate in our galaxy, and other similar galaxies is now to a good approximation equal to the average rate over the lifetime of the galaxy, which is about 10^{10} years. The time scales for the processes involved are small, compared to this lifetime, 10^5 years for the collapse to new stars and 10^7 years for the time between formation and supernova of a massive stars; to maintain this non-equilibrium configuration stably over so many dynamical times there must be feedback processes that control the rates of formation of clouds and stars.

Further *a priori* evidence for the existence of processes governing these rates

---

[5] Evidence, for example, is that massive stars tend to form in clusters, with the most massive in each cluster often formed last.

is that the rate by which material is converted from the interstellar medium into stars, which is about $3-5\ M_{solar}$ per year, matches well the rate at which matter is returned from stars to the medium through supernova and stellar winds, which is estimated to be at least $1-2\ M_{solar}$ per year. To astrophysical accuracy, these numbers could be equal, but even if they are not, they are close enough that some explanation is needed. Related to this is the fact that although star formation has been going on for 10^{10} years, it is the case that in the midplane where these processes take place, fully half of the matter is presently in gas and dust.

In thinking about these things it is important to emphasize that the galactic disk seems to be an open system. Old stars evaporate off of the midplane at a constant rate, as their encounters with other stars give them velocities perpendicular to the plane. Further, it may be the case that new gas continually or intermittently enters the system, either by infall onto the disk or by inflow into the star forming regions from the gaseous disk that seems to extend quite a bit out of the visible galaxy.

Another kind of evidence that there ought to be mechanisms that control the rate of star formation in spiral galaxies is that there are galaxies where this apparently does not happen. Little or no star formation is taking place in elliptical galaxies; they contain no dust and what little gas they have is heated to the point where further star formation seems unlikely. At the other end of the scale are the so called star burst galaxies that are forming stars at rates that are not sustainable for long periods. Many of these are small or dwarf galaxies, which seem to be found in either a star burst mode or in a quiet mode with little star formation. The evidence is then that to achieve the steady, sustainable rates of star formation that are seen in spiral disks requires a system of a certain size.

All of this evidence suggests that there must be mechanisms that explain how a spiral disk achieves a steady rate of star formation. Several hypotheses have been made about such mechanisms. I will describe a few of them.

Parravano has proposed a feedback mechanism that regulates the rate of star formation by controlling the rate of condensation of the giant molecular clouds [21]. The idea is that the mechanism maintains the interstellar medium at the critical temperature at which the giant molecular clouds may exist in equilibrium with the warm ambient gas. This critical temperature depends on the pressure and other factors such as the amount of dust and there is evidence that the interstellar media of a large number of galaxies are near it [21]. A mechanism that would maintain the medium at this critical point works as follows. As the temperature falls below the critical point giant molecular clouds condense, which leads (in combination with other factors) to the formation of new massive stars. The ultraviolet light from these stars heats the medium, raising its temperature above the transition. This cuts off the formation of new clouds, and hence new stars. But after about 10^7 years this leads to a decrease in the ultraviolet radiation, leading to the cooling of the gas below the critical point, and so on.

What is interesting about this mechanism is that it functions on scales larger than individual clouds, tieing the rate of star formation to the rate of cloud condensation. If there is a mechanism to guarantee massive star formation given the existence of giant molecular clouds it can explain why this process may continue at a steady rate as long as the pressure is sufficient for there to be a critical temperature. It is then interesting that the supernovae can both provide a mechanism for star formation given the presence of enough molecular clouds, through the self-propagating star formation, and provide the energy which pressurizes the interstellar medium.

Let us then assume that there is a mechanism such as Paravanno proposes that keeps the medium critical so new molecular clouds are condensing from the ambient gas at a steady rate. There will then be a steady rate of star formation. We then want to ask more detailed questions about the geometry of the star forming regions. This kind of question was addressed by the Seiden-Schulman-Gerola model. In this model the process of self-propagating star formation is described as a percolation process, which is then modeled by a cellular automata. The model is simple, in some ways very like the game of life, put on a rotating lattice.

The model is constructed by dividing the disk into rings, each of which is divided into a number of cells. The disks rotate differentially, at the same linear speed, in order to match the flat "rotation curves" that are generally observed in disk galaxies. The model evolves in discrete time steps according to simple local rules. Each time step is about 10^7 years, which is the typical time between the birth of a massive star and its destruction in a supernova. At each step each cell may be on or off, which represent whether star formation is occurring or not. Each cell also has associated with it a quantity of gas, which is distributed among two states, which correspond to the warm ambient gas and the cold clouds. It is assumed that in each time step in which star formation does not occur in a cell, a certain proportion of its gas condenses from warm to cold clouds. But during a step in which star formation occurs, all the gas is heated and returned to the warm state.

The rule of evolution is then stochastic. There is a small spontaneous probability for star formation to occur and an induced probability which is proportional to the number of neighboring cells in which star formation occurred in the last time step multiplied by the amount of cold gas in the cell.

The model has three parameters: The radius of the galaxy, which gives the number of rings, the rotation velocity and the rate at which cold clouds condense from warm gas. The latter gives a "refractory period" over which star formation is unlikely to repeat in the same cell due to there being insufficient cold clouds for stars to form. Over a wide range of parameters, the model seems to show what might be called self-organized critical behavior, in which star formation occurs at a steady rate. In this state, the star forming regions make spiral patterns that continually form and dissolve. Furthermore, given a suitable choice of parameters, these resemble rather well the patterns seen in some spiral galaxies.

It is possible to interpret the model in the following way: the dynamics of the gas is providing a feedback mechanism which is tuning the system close to the critical point of a percolation problem. Indeed, one may simplify the model by eliminating the gas. Then the induced probability for star formation to occur in a region is simply proportional to a parameter times the number of neighbors in which stars are forming. In this case the system is a directed percolation problem in $2 + 1$ dimensions. There is a percolation phase transition and to get spiral patterns that continually form and a constant rate of star formation one must tune that parameter near the critical point for the transition. What the gas dynamics seems to do is to eliminate the need for adjustment of a parameter by keeping the system in the critical state by a feedback process.

Some astronomers have criticized this model for oversimplifying the real phenomena and also for being unable to describe certain kinds of spiral galaxies. Their criticism is in part correct, but in a way also misses the point. It is true that important phenomena are neglected in this model, for example the gravitational dynamics of the stars and the medium are completely ignored. It seems that in some galaxies this is justified. In these, the spiral patterns are seen only in the distribution of star forming regions, and hence are observed only in the blue light coming from the bright young stars, and not in the red light coming from the old stars. In these the spiral patterns tend to be fluffy or "flocculent", and it is these kinds of patters that seem to be well modeled by the Seiden-Schulman-Gerola model.

In the older models favored by some astronomers, the opposite simplification is made. The gravitational interactions among the stars are modeled, and the energetics of star formation and supernova, as well as the processes governing the conversion of material between stars and the interstellar medium are ignored. In these models one sees that density waves can be excited in the distribution of stars in the disk. These can catalyze star formation, because a giant molecular cloud can be perturbed significantly by falling into the deeper gravitational potential of a passing density wave. According to such a model, the spiral patterns should be seen both in the new stars, tracing the star forming regions, and in the old stars, tracing the density wave. Furthermore, in such models the density waves, and hence the spiral patterns can show bilateral symmetry, so one can have strongly symmetric spiral arms, and not just a kind of spiralling fluffy pattern.

Galaxies of this kind, in which the gravitational dynamics seem important, are seen. Clearly these are not going to be modeled by the Seiden-Schulman-Gerola model. However, the density wave models have problems of their own. The density waves must be excited, either by an outside perturbation such as a passing galaxy or by an asymmetric field such as might be generated by the galaxy itself. Such asymmetric structures are seen, they are usually in the form of bars. However, spiral structures are seen in galaxies that are without bars and are also apparently isolated far from other galaxies. In these cases there is a problem as the density waves are damped, and will die out after a few rotations.

Clearly what is needed are models that contain both elements. Although it is not the most elegant possibility, it is hard to avoid the conclusion that there are some galaxies in which the spiral structure traces density waves and there are other galaxies in which the spiral structure is not traced in the density, and is more a result of self-propagating star formation near a percolation phase transition. This point of view has been advocated by Elmegreen, who, with Thomasson has constructed such a hybrid model [27]. In this model, the gravitational dynamics of the stars and the energetics of star formation and the interaction of stars and clouds are included. This model seems, for appropriate choices of parameters, to be able to describe either kind of spiral structure.

At the same time, while it describes a wider range of phenomena, the Elmegreen-Thomasson model requires that certain parameters that describe the energy balance between the populations of stars and clouds be tuned so that a constant amount of energy is maintained in the system. This tuning of rates to maintain the energy balance is presumably accomplished in nature by the kind of feedback mechanisms that are modeled by the Paravanno hypothesis and the Seiden-Schulman-Gerola model. Thus, while it may be a satisfactory model of spiral structure, the Elmegreen-Thomasson model still does not represent a complete model of the energetics of the disks of spiral galaxies.

But my point here is not to debate whose model is better but to make the point that it may be useful to describe the disks of spiral galaxies as self-organized critical systems. Let me then end this section by summarizing the reasons why it seems reasonable to think of the disks of spiral galaxies as self-organized critical systems.

That they are critical is to be seen from:

- The simultaneous presence of several different phases in the interstellar medium with very different densities, temperatures and compositions, again over very long time scales.
- The evidence of Paravanno that many galaxies are near the transition point for simultaneous existence of warm gas and cold clouds.
- The suggestions that the distribution of densities in the cold molecular clouds is scale invariant.
- The apparent long ranged order in the spiral structure, which, together with the mechanism of self-propagating star formation suggests a percolation system near a critical point.

The evidence that they are self-organized comes first of all from the evident fact that as galaxies are isolated, any critical behavior that is widely seen must be arrived at spontaneously, without the need for tuning of external parameters. Besides this, there is evidence from,

- Constant star formation rates, over time scales very long compared to the dynamical time scale.
- An approximate balance between the rates of flow of matter in each direction between stars and the medium in the disk, despite the possibility of loss of stars by evaporation to the halo and inflow and infall of gas to the disk.
- The success of the hypotheses of Paravanno, Seiden, Schulman and Gerola and others that there are feedback mechanisms which maintain the gas in the disk in a critical state with a constant rate of formation of cold clouds, which matches their rate of destruction.

3 What is the large scale organization of the universe?

Probably the key cosmological problem at present is that of the formation of the galaxies and their large scale organization. The amount of data we have about the history and organization of the universe on scales larger than galaxies is increasing quickly; and the theories have consequently been evolving very rapidly in this domain.

Given the apparent usefulness of conceiving of a galaxy as a self-organized non-equilibrium system, it is natural to ask if new concepts from non-equilibrium statistical physics such as self-organized criticality might be useful for understanding how structures on still larger scales emerged. There are three reasons, *a priori* for imagining that this might be the case. First, the processes that formed the galaxies and their large scale organization occurred at earlier times when the universe was on average denser and hotter. It is then natural to ask if non-equilibrium processes such as those we see dominating the process of star formation might have played a role in some denser era in the formation of galaxies. To put this another way, we now understand galaxies to be dynamical systems, in which supernova and other energetic processes play a dominant role. It is then natural to ask whether such processes might have played a role in their formulation. Second, there are senses in which the distribution of galaxies and clouds of gas are approximately scale invariant, which may suggest the study of galaxy formation as an example of a critical system. Third, there is a sense in which all gravitationally bound systems are intrinsically out of thermal equilibrium.

I would like to briefly expand on this last point. While gravitationally bound systems may spend long periods of time in quasi equilibrium configurations, they do not reach true equilibrium states, characterized by maximal entropy[6]. The reason is that they have practically inexhaustible sources of energy, coming from gravitational binding energy of subsystems. A subsystem can always become more deeply gravitationally bound, freeing energy to other parts of the system. A consequence of this is that all large gravitationally bound systems are characterized by a flow of energy at some rate from gravitational energy to heat or kinetic energy. The question is only the rate. When coupled with another source of energy, nuclear energy, gravitationally bound systems such as galaxies can maintain significant flows of energy for cosmological time scales.

It is significant that what characterizes self-organized critical systems is that they are kept out of equilibrium by steady flows of energy through them from a source to a sink. Large gravitationally bound systems do this naturally. It is then interesting to speculate that all large gravitationally bound systems may, to one extent or another, be thought of as self-organized critical systems. This description apparently is suitable in the case of spiral galaxies, it is then interesting to ask if the flows of energy in other systems and on other scales is significant enough to play a role through mechanisms of self-organization.

---

[6] There actually is available an equilibrium state for any isolated gravitationally bound system, it is the black hole containing the total mass and angular momentum of the system. Fortunately, the time required for most astrophysical systems to reach this state is much larger than the Hubble time.

The evidence we have presently for the large scale organization of the universe comes from many sources. The most important methods have been 1) catalogues of galaxy redshifts; 2) absorption lines in quasarspectra, 3) the cosmic black body radiation, 4) studies of the distribution of hot ionized gas in clusters of galaxies, by measurements of the x-rays they give off and 5) measurements of large scale velocity flows, by careful combinations of distance and redshift measurements. Together these give a detailed picture of the organization of matter in the universe, and the amount of data available is expected to increase dramatically over the next years. A theory of cosmology must account for all of these data by a detailed description of the history of the universe that begins perhaps 10^{-43} after "the big bang" and runs to the present. It is a tall order, and it must be said that the existing hypotheses do not do badly at the present time. But there are issues that suggest that the present picture is incomplete; I sketch here a few of them.

3.1 Dark matter and the issue of Ω

Any understanding of the large scale organization of matter in the universe must take into account the evidence that at least eighty percent of it is not visible. The strongest evidence for this comes from the rotation curves of galaxies, which leads to the conclusion that most galaxies are surrounded by large halos of non-luminous matter, with between five to ten times as much mass as is present in visible stars, gas and dust [43]. In units of Ω, where $\Omega = 1$ would be exactly enough matter to close the universe, the visible matter in galaxies is about $\Omega_{observed} = .01$, whereas the total gravitational mass in galaxies is roughly ten times larger.

Other evidence comes from careful studies of clusters of galaxies. Measurements of X-rays from large clouds of ionized hydrogen surrounding the galaxies lead to a conclusion that there is no more than about ten times more gravitating matter than is contained in the observed baryons. This, together with the bounds coming from nucleosynethesis, which is $\Omega_B h_{50}^2 = 0.05 \pm 0.01$, leads to the conclusion that that $\Omega = .1 - .2$ [1].

The question is then whether there might be still additional non-luminous matter, clustered on still larger scales, that could increase Ω, perhaps to unity. While the evidence for a low value of Ω in the range $.1 - .2$ seems to be increasing [1], we may note that there are observations of large scale flows of matter that, given certain theoretical assumptions, point to a larger value [44]. A number of other observational issues bear on this question including the value of the Hubble constant, the question of the mass of the neutrino, the age of the oldest stars in globular clusters, and the abundances of rare primordial elements. It seems likely that there will be significant progress on all of these questions, so that we may hope within a decade or two for a sharp resolution of the value of Ω.

There is a strong theoretical reason for a value of $\Omega = 1$, which is that it seems to be required by all natural inflationary scenarios. It is possible to invent inflationary models for which $\Omega < 1$, but these require an additional tunings of a certain parameter [28]. Theorists may disagree on the extent to which this is a cause for worry, as there are already at least two fine tunings that must be done for any inflationary scenario to work, and to yield a reasonable spectrum of fluctuations, first of the cosmological constant and second of the self-coupling of the "inflaton" field. This is not to say that fine tuning is not a problem, but only that if inflation is to be in the end accepted we must uncover a natural mechanism to accomplish these fine tunings; if such a mechanism is discovered it may as well be able to fine tune the inflationary mechanism so that $\Omega < 1$.

If Ω does fall in the range $.1 - .2$ favored by most current observations, it may free theory from having to provide an exotic non-baryonic particle for the dark matter. Given the apparent failure of pure hot dark matter models, we know the non-baryonic dark matter cannot be only massive neutrinos; so any theory that demands Ω to be equal to unity requires that we postulate that the universe is dominated by a kind of matter for which we have no observational evidence.

On the other hand, if $\Omega \neq 1$ then the universe has an intrinsic scale written into it's initial conditions, which is greater than its present age. Assuming that the initial conditions are set at some early time by the action of physical processes involving quantum gravity or grand unification, leads then to a puzzle, for we must ask how physical processes involving time scales of 10^{-43} seconds could be fine tuned in a way that implicitly involves a time scale of 10^{17} seconds.

3.2 Quasar absorption lines and the universe at earlier times

A window into the distribution of matter in the universe of increasing importance is the analysis of the absorption lines of quasars. Many quasar, have redshifts in the range of $z = 2 - 5$, and were thus active when the universe was significantly smaller. It turns out that whenever the light from a quasar passed through a sizable enough cloud of gas on its way to us we see absorption lines at the appropriate redshift. Most of these lines are due to the Lyman alpha transition in hydrogen, and some are produced by heavy elements such as carbon and magnesium.

More than 150 quasar spectra have been studied, and each of them contains on the order of 100 lines, so that there are reasonable statistics about the distribution of clouds of gas between them and here.

The basic results seen in these observations are the following [38],

- There is little or no unionized hydrogen between the clouds. For example, at a redshift of 2.26, the ratio of unionized hydrogen seen to the average matter density is less than 10^{18} [38]. This most likely means that the intergalactic medium is ionized, up to at least a red shift of $z = 5$. The source of the energy to ionized the medium is unknown; this is itself an important problem. Possible candidates are the quasars themselves, an early generation of supernovas or massive stars. There are also exotic possibilities, such as the decay of massive neutrinos.
- From the Lyman alpha absorption lines one may measure the column density of neutral hydrogen in each cloud, which is the number of atoms per square

centimeter in the line of sight of the quasar through it. Remarkably, over a range of at least nine orders of magnitude, from 10^{13} to 10^{21} $atoms/cm^2$, the distribution of clouds at a given redshift satisfy a power law distribution in column density σ;

$$n(\sigma) \approx \sigma^{-\gamma} \tag{55}$$

with $\gamma = 1.67 \pm .19$ [38].

At the high end, the column densities are comparable with those through the central region of the disk of a spiral galaxy. It is intriguing that these are seen to fit into a single power law with much more diffuse column densities. Because we are seeing through a random line of sight through each cloud, the distribution of column densities may be a combination of two factors, the distribution of densities within a given cloud and the distribution of masses of the clouds themselves. One may make a number of hypotheses about both. However, whatever combination of these factors determines the power γ, the fact that there is one power that ranges from the densities of galaxies down suggests that one mechanism must be responsible for the formation of the galaxies and the clouds seen in the absorption lines. This is particularly impressive as there are so many orders of magnitude involved[7]

 - One hypothesis that may be explored is that the galaxies are surrounded by large diffuse clouds of gas, that are in approximate hydrostatic equilibrium, and so have densities that fall off like r^{-2} as we go from the center. There is increasing evidence for such a picture in the study of correlations between the denser absorption lines and actual galaxies near to the line of sight of a quasar [40].

Very recent observations suggest find that, at least for low redshifts, if a galaxy is within $40kpc$ of the line of sight there is almost always an absorption line in hydrogen with a column density of at least $10^{15}/atoms/cm^2$ and vice versa [40]. This suggests a picture in which many galaxies are surrounded by spherical clouds of hydrogen and other gases which extend out to at least $40kpc$. These clouds, are often seen also in carbon and magnesium, so that it appears that they have been enriched by the action of supernovae. It is then very interesting to know whether this enrichment came from supernovas at an earlier time, took place during the formation of the galaxy itself, or, on the other hand, is the result of outflow from the galaxies themselves.

It is then interesting to try to imagine that these clouds and the galaxies they contain are single systems, with significant exchanges of matter between then, perhaps in both directions. One may wonder, for example, whether the observed constant star formation rates of spiral galaxies are related to the rates at which gas falls from the surrounding clouds onto the disk.

 - Finally, the quasar absorption spectra give very good probes of the distribution of matter at high redshift. One intriguing result is that at very high redshift $z > 4$ there are about four times more of the densest absorption lines than would be given by the present day galaxies. The interpretation of

---

[7] There is also the possibility of a break in the distribution, so that the distribution has slightly different powers at high and low column densities [39].

this is problematic; it may be that many clouds never formed into galaxies, or it may be that the clouds have contracted significantly since then.

However, we must keep in mind another interesting thing, which is that there is evidence that the properties of large galaxies have not changed very much since redshifts of $2 - 3$, which is on the order of ten billion years [41]. Before that time, energetic processes, such as those that fuel quasars, were much more common then they are presently, however there seems to be a sharp decrease in the numbers of quasars seen after red shifts around 2, suggesting that large normal galaxies have since that time established a kind of equilibrium[8]. The evidence we mentioned above agrees with this picture, suggesting that normal spiral galaxies have a constant rate of star formation over most of the time since their formation. This, together with the evidence I summarized above, suggests that it might be fruitful to understand the galaxies and their surrounding gas clouds as single stable far from equilibrium systems.

3.3 The issue of homogeneity on very large scales

There is a final issue I should mention, which has been the subjection of discussion among statistical physicists. This is the question of the large scale homogeneity of the universe.

Since Einstein and DeSitter, cosmological models have always been based on the Cosmological Principle, which assumes that we live in a typical place in the universe. It is also observed that to very high precision, the universe is isotropic to a very good approximation. This can be seen in the $COBE$ radiation, which is isotropic up to a part in 10^5. As the radiation has passed through the gravitational potential of matter on its way here, this puts limits on the anisotropy of the distribution of matter from redshifts of 1000 to the present. Counts of galaxies, or radio sources also show impressive evidence of isotropy [43].

If our view of the universe was perfectly isotropic, and it were so, by the Cosmological principle, around every point, then we would have to conclude that it was perfectly homogeneous. The difficulty is that it is neither perfectly homogeneous nor perfectly isotropic, which raises the issue of how it is to be described.

The simplest assumption is that the departures from homogeneity decrease at large scales, so that there is some scale λ_h above which the universe may be satisfactorily described as homogeneous[9]. This assumption is usually made by

---

[8] However, it should also be mentioned that the observations indicate that the much smaller "dwarf" galaxies have evolved a great deal since redshifts of 2, there seem to have at that time been more of them than there are now, especially the "blue" ones, in which a lot of star formation is going on [42].

[9] At least up to some larger scale, we may note that no cosmological observation is able to constrain the homogeneity of the universe on scales larger than the distance to our horizon, so that it is perfectly possible that the universe is very inhomogeneous on some much larger scale. This possibility is taken advantage of in the inflationary models, which describe the universe as a single bubble that inflated. The bubble does have walls, even if we can't see them.

astronomers, and so far there is no evidence against it.

The difficulty is that the large scale surveys of the galaxies, which map the distribution of matter, show so far structures that are as large as the scales of the surveys [5]. Furthermore, at least up to the scale of clusters of galaxies, the distribution of matter is approximately scale invariant. This means that one of two things must happen. As the surveys increase in depth, the scale λ_h must be discovered, or structures must continue to be found on every scale up to the horizon. It has been thus suggested that perhaps the standard assumption is wrong, and the universe has a fractal (or multifractal) structure on all scales up to the horizon []. The difficulty with this picture is that such a distribution should also agree with the isotropy seen in both the counts of galaxies and radio sources and in the microwave background, as well as with the Cosmological Principle []. The question is whether there can be a distribution that shows inhomogeneities on arbitrarily large scales that is in agreement with this.

A related question is how to describe a universe that is inhomogeneous over a large range of scales in general relativity. Clearly it will not do to work solution by solution, what is needed is something like a renormalization group treatement, that lets us think in terms of coupling between modes on different scales. The tricky thing is how to to do this in a way that is generally covariant, since the metric that measures scales is dynamical. A very interesting step in this direction has been taken by Carfora and Piotrkowska [45]. Even if there is a scale above which the universe is homogeneous to a good approximation, there are corrections to the equations that describe the expansion of the universe coming from averaging over the fluctuations at smaller scales. An important, and presently unresolved question, is to determine if these corrections may lead to significant modifications i the age of the universe [45].

4 The problem of the origin of the large scale structure

We have been discussing the evidence that tells us how the universe is organized on large scales. Now I would like to turn to the question of what is understood about how that structure has arisen.

The first thing that must be said is that astronomers have developed numerical models of the evolution of structure in the universe that seem to go quite far towards explaining features of the observed distribution of galaxies. I would like to begin this discussion by summarizing how these models work [44, 43, 35].

The models take as inputs certain assumptions about the conditions of the universe at decoupling. These begin with a specification of the basic cosmological parameters, such as Ω, the value of the cosmological constant and the amount of dark and baryonic matter present. Because of the isotropy of the present universe, and the fact that it works so successfully, the universe is always assumed to be homogeneous and isotropic, with an initial spectrum of perturbations whose amplitudes are small (on the order of 10^{-5}) on all scales.

To this picture one then must add several assumptions. First, one must choose between two general types of initial perturbations. Adiabatic perturbations are those for which the baryonic and photon densities fluctuate together, so that the observed temperature fluctuations observed in the COBE signal trace also fluctuations in the density of baryons. Another choice is to take what are called "isocurvature" perturbations, in which there may be larger fluctuations in the density of baryons, which are, however, not reflected in the distribution of temperatures, because the distribution of thermal photons does not trace the distribution of matter. The first is better studied, but both are reasonable possibilities.

A very important assumption that must be made is the spectrum of initial fluctuations. The assumption that is most often made is that the initial spectrum of fluctuations is approximately scale invariant, this is the simplest possibility and was proposed some time ago by Harrison and Zeldovich. It is also what is predicted by inflation. The amplitude of the spectrum may then be normalized by the COBE measurement.

In the near future the measurements of the black body spectrum are expected to be very much improved, so that the initial spectrum of fluctuations will, in the adiabatic case, be largely constrained by observation. Of course, this will not constrain the isocurvature models as much, as by assumption they take the initial perturbations in the baryons to be decoupled from those of the photons.

The last assumption that is made in the construction of these models is the type of dark matter present. These may be of several kinds: dark matter may be hot or cold, depending on whether their masses are small or large compared to the cosmic background temperature, it may also be baryonic, in the case it consists, for example of black holes, or non-baryonic, as in the case of neutrinos or hypothesized particles such as axions.

Given these choices, the numerical simulations have been able to show how the perturbations grow, leading to the structures we see today [44]. While there are important differences between the models based on different assumptions, a variety of assumptions are known to lead to structures very much like those we see today. Very roughly, in all of them perturbations grow through a long linear phase, from their initially small values to values of order one. After that, non-linear processes involving both gravitational binding and hydrodynamics effects take over, leading more or less quickly to the formation of galaxies and clusters of galaxies.

It must be emphasized that it is nontrivial that the models work at all, given the simplicity of the assumptions made. Given that the spectrum of perturbations at decoupling is constrained, in the adiabatic case, to such a small value by the COBE data, and given that the age of the universe is also constrained, to within a factor of two, it might very well have been the case that structure forms at too slow of a rate to explain what is seen at the present time.

There are, however, a number of places in the picture in which complementary or alternative points of view may play a role. These include particularly the role of non-linear processes in structure formation.

4.1 Understanding the non-linear stages of galaxy formation

According to the standard models of structure formation, once the perturbations in the distribution of mass and baryons become of order one, non-linear processes take over, leading to the formation of the present day structures. While there is a good analytic description of the linear regime, there is no correspondingly successful treatment of the non-linear regime besides the large scale numerical simulations.

There are several possible indications that self-organized critical phenomena may play some role in this non-linear regime.

- The structures which are formed are scale invariant, and governed by power law distributions at least up to the scales of clusters of galaxies. Furthermore, as I mentioned above, the structure of clouds of gas, back even to red shifts of 4 − 5 follows a power law over 10 orders of magnitude, as seen in the distribution of quasar absorption lines. Thus, irrespective of the question of the large scale organization, it is clear that the distribution of galaxies and gas may be characterized as fractal over many orders of magnitude.
- There are suggestions that several features of the final distribution of galaxies and mass may be independent of the detailed assumptions that go into the large scale simulations. These may include the powers that govern the distributions of galaxies. If so, this suggests that there may be simpler statistical arguments for some features of the observed distributions.
- There is a rather simple model of hierarchical structure formation due to gravitational binding in an expanding universe that does agree to some extent with both the observed distributions and the results of the numerical simulations. This is the Press-Schecter formulation [46], which I would like to briefly describe

In a very interesting paper, Press and Schecter described both analytic and computer models of a collection of gravitating particles in an expanding universe. The particles originally have equal masses and are randomly distributed. At certain intervals, as the universe expands, a test is applied to identify clusters that are gravitationally bound. In their computer simulations Press and Schecter considered spherically regions which were overdense by a factor of 10 to be bound. Those clusters that are bound at each step are replaced by single particles with a mass which is the sum of the masses of its members.

This processes is iterated and it is found that after a time an approximately scale invariant distribution of masses develops which has the form

$$n(M) \approx \frac{1}{M^{1.5}} \qquad (56)$$

for small masses, times a high mass cutoff e^{-kM/R^2} that scales as R, the scale factor of the universe.

Press and Schecter found that the same approximate scale invariant distribution resulted from their model, given different kinds of initial distributions of the particles. They also gave a simple analytic derivation of the scaling law. Finally, they were able to compare the predictions of this model with observation and they found that the distribution of luminosities (which scale with mass) of galaxies in the Como cluster scale with an approximate power law. Since that work was done, both observations and numerical simulations of the distribution of galaxies in clusters has tended to support this simple picture [47].

We may note that in the formation of an apparently universal scale invariant distribution from different initial conditions, the Press-Schecter model might be described as a very simple kind of self-organized critical system.

4.2 Possibilities for early structure formation

Finally, even though the simulations of galaxy formation based on the standard dark matter scenarios do seem to work reasonably well, there is still the possibility that they are based on assumptions that may turn out to be incorrect. Especially given that some features of the observed distribution of galaxies may be produced by non-linear effects that wash out some of the information about the initial conditions, we must keep open the possibility that more detailed observations, especially at higher redshifts, may turn out to be inconsistent with these models. It may then be useful to consider other kinds of models which may account for the observed structure[10] While this might be considered a higher risk activity than the others on my list, it may be motivated by consideration of the fact that there is a certain lack of economy in the assumptions that must be made in the standard models. At present, the nature and properties of both the dark matter and the initial perturbations are essentially ad hoc, and can be manipulated to yield results consistent with observations. It would certainly be more elegant to have a theory in which there was not so much room to introduce ad hoc elements.

One might then dream that a scenario for cosmology could be made to work in which nonlinear processes play a role much earlier in the history of the universe, acting near or just after decoupling to produce the spectrum of fluctuations that become the large scale structure[11]. In such a picture, the slow growth of primordial fluctuations after decoupling would be replaced by a picture in which non-equilibrium processes act at very high redshifts of 500-50 to produce a spectrum of fluctuations in the distributions of matter and baryons that might be largely independent of whatever initial perturbations are present at decoupling.

We may note that the fact that the isocurvature models are consistent with present knowledge means that it may be that the perturbations seen in the black body radiation do not trace the perturbations in the matter (although there are

---

[10] Many suggestions have been proposed that depart in small or large ways from the standard structure formation scenario I sketched here. Several involve explosions or other energetic events in the early universe [31, 32, 33]. Others interesting proposals involve a low density, $\Omega \approx .1 - .2$ universe [35, 36, 37].

[11] Two very interesting attempts to model structure formation in the distribution of galaxies are by Chen and Bak [48] and Schulman and Seiden [49].

limits based on the motion of the light through the inhomogeneous gravitational fields of matter since decoupling.)

Such a scenario could take advantage of the fact that at redshifts of around 100 − 200 the conditions of the universe as a whole are not that dissimilar, in density and temperature, from those which characterizes the interstellar medium of the disks of spirals galaxies. It is then possible that non-linear processes that are analogous to those that are responsible for the spiral structure in galaxies might act to form structure at these earlier times. The amount of time that such processes would have to act is limited by the expansion speed at that time to at most several hundred million years. But this is one to two orders of magnitude longer than the lifetimes of the massive stars, making it possible that processes in which massive stars are formed and inject a great deal of energy into the medium could produce significant structure during this time.

There are in fact some reasons to believe that there was an era of star formation before the formation of the present day galaxies, coming from the need to explain both the fact that some enrichment is seen even in very old clouds of gas and the fact that the intergalactic medium is ionized back to at least redshifts of around 5. At the same time, such a scenario would have to be limited by the requirement that not too many heavier elements were produced [50].

One may also try to understand if, in the context of such a scenario, it is possible if the dark matter could be formed as a consequence of such early processes of structure formation, rather than having to be posited independently. One way this might work is if a very early era of star formation processes produced large numbers of neutron stars or black holes, which made up some or all of the dark matter that then dominates the structure and formation of the galaxies at later redshifts from about 20 to the present. A dark matter scenario in which the non-luminous matter consists of black holes which are formed in the same processes that make the the galaxies and stars might be more parsimonious than the standard scenarios in which the dark matter is put in by hand to account for the structure formation.

At the same time, the possibilities for such early structure formation processes are constrained from several sides, including limits on the numbers of black holes coming from MACHO searches and other observations.

5 The problem of the parameters in particle physics and cosmology

In the introduction I stressed that many of the key problems in cosmology rest on problems of fine tuning involving the parameters of particle physics and cosmology. It is indeed, not an exaggeration to say that the fact that we live in a world which is large, complex, out of thermal equilibrium and full of a large variety of phenomena is a consequence of the parameters being tuned to special values. There are two kinds of such fine tuning problems. The first involve issues of hierarchies, in which parameters have improbably small values, such as in the case of the values of the proton or electron mass, in Planck units. The other class involves cases in which structures of a certain kind would not exist if a parameter were to take values different from its present ones, by less than an order of magnitude. Examples of this are the proton-neutron mass difference, the electron mass, or the strength of the fundamental electric charge: increases in any of these separately, by factors less than ten would result in a world with no nuclear bound states, and hence no nuclear or atomic physics.

There are two responses that have traditionally been made to the problem of the values of the parameters of particle physics, in the light of this situation. The first is to hold to the faith that there is a unique fundamental theory that after a pattern of spontaneous symmetry breaking and, perhaps, dimensional reduction, will have a ground state whose low energy excitations will match the pattern of elementary particles and forces that we see.

As the existence of such a theory has been taken to be almost axiomatic by many theoretical physicists, let me spend a moment to suggest its likelihood is not so obvious. First, there is no evidence for the existence of such a theory, at least at the perturbative level. In the last ten years we have learned that there are very large numbers of perturbative string theories, which give equally consistent unifications of the strong weak and electromagnetic interactions with gravity, but in different dimensions, with different low energy physics. It may be that there is one non-perturbative string theory and these perturbative theories are all descriptions of excitations of its many ground states. But there seems, at this point, little evidence for this[12]. Instead, it may be observed that there seems to be a logic under which, the more disparate fields are incorporated into a unification by a gauge symmetry, the more it is the case that the properties of the low energy excitations depend on a choice of the ground state of the system. Thus, in theories which incorporate the Higgs mechanism, the masses of the low lying states depend on coupling to a condensate. If there are many degenerate or nearly degenerate ground states, with different properties, then it may be said that the masses and couplings of the light particles are determined cosmologically, as the ground state may depend on the history or configuration of the universe as a whole.

Thus, in a certain sense the assumption that the properties of the elementary particles are independent of the state and history of the universe as a whole is breaking down. To the extent that this happens, elementary particle physics and cosmology become interwoven, and the Newtonian conception that a particle in a universe that contained it alone would be just like a particle in our universe becomes untenable.

Certainly the inflationary models work in this way, as the spectrum of light

---

[12] There is recent evidence that the moduli spaces associated with the diffferent Kalabi-Yau compactifications may be connected to each other through singular configuations that may represent critical points in the parameter space where certain fields condense [51]. It is then possible that there is a single non-perturbative ground state in which the quantum state is spread out over this single extended moduli space. But, it is also presently a possibility that what is being described is a very large family of degenerate ground states, which are able to tunnel to each other by going through the singular configurations.

particles is different before and after the phase transition that simultaneously determines the large scale properties of the universe. This also, I would argue is one lesson we have learned from string theory in the last ten years; whether or not there is a nonperturbative string theory whose vacuum states they describe, the fact of these many different perturbative theories means that consistency alone does likely not govern the choice of the phenomenology of the light particles.

If the standard model of particle physics is not to follow uniquely from demanding only consistency, there must be another kind of principle which picks out which, of the many equally consistent worlds, is the one we find ourselves in. Because of the coupling between the selection of a ground state and the history of the universe, this means that the hard questions in elementary particle physics are likely closely connected with the hard questions in cosmology. It is then remarkable that in both cases these hard problems involve understanding unnatural choices of the values of parameters.

The second response that has been given to this situation is the anthropic principle. This states (in what is called its weak form) that the choices of parameters that lead to our world may be picked out by noticing that it is among a rather narrow range in which intelligent life can exist.

Now, as stated this is undoubtably true, indeed, it is an aspect of the fact I have already stressed, which is that with most choices of parameters a world would not have the complexity of ours. The question is whether this observation can be made into an explanatory principle. Rather then deal with this philosophical question at length (again, this is done elsewhere), I would like here only to ask if it is possible to do better. That is, is it possible that there might be a mechanism that could explain how the parameters were chosen that accounts for the fact that the actual values selected lead to a world with the complexity of ours.

I know of one such theory, that does seem to yield non-trivial testable predictions. I will briefly describe it here, for more details the reader may consult references [52, 53, 54].

5.1 Cosmological natural selection

This theory comes from two simple conjectures about quantum gravity, neither of which is really new. The first is that there are no final singularities in nature, instead, due to quantum effects that are ignored in the singularity theorems, singularities inside of black holes, and final singularities of cosmological spacetimes are replaced by "bounces" as a result of which the collapsing matter reverses its collapse and begins to expand again. This is an old idea that goes back at least to Lemitre's "Phoenix universe" and has been discussed by Wheeler [55] and others. Recently, plausible scenarios for how this might occur have been discussed in the context of string theory [56] and inflationary models [57]. However, to get definite physical predictions, as I will show we need know nothing about how this happens except simply that each black hole and cosmological singularity turns, one for one, into a new expanding region of space and time.

The second conjecture I will make is that when this happens the parameters that govern the low energy physics and large scale cosmology of the new *region* differ from the parameters of the one in which the collapse took place by small random fluctuations. This is also an old idea, which was proposed, in the case of cosmological singularities by John Wheeler, who called it "the reprocessing of the universe." I need to add to it only the assumption that the changes are small. Of course, I will have to say what I mean by small, I will do this in a moment.

There are also plausible homes for such an idea in grand unified theories or string theories, as in each case there are large families of vacua, which correspond to different compactifications and symmetry breakings. It is quite plausible that a violent, Planck scale event like the bounce may force the vacuum to jump or tunnel from one ground state into a nearby one, leading, after the region has expanded into a large universe, to a small change in the parameters of low energy physics.

However, again, while it may be important to develop such theories, the predictions of this theory are independent of the details.

Our universe has at least 10^{18} black holes in it, so that given these assumptions we are dealing with a universe with an enormous number of regions, in which we find a distribution of different parameters of low energy physics and cosmology.

However, given only these two postulates, we may make non-trivial predictions about the parameters that characterize our world if we add only one more assumption, which is the Copernican postulate that our world must be a typical member of this ensemble. We can then make predictions about our world if there are statistical predictions that can be made about the properties of randomly chosen members of this ensemble.

We can do this because this theory is isomorphic to models of biological evolution, in which natural selection is described in terms of the evolution of probability distributions on fitness landscapes. As a result there is a natural mechanism of cosmological self- organization, that is formally analogous to biological natural selection.

It goes like this. We may consider the space of parameters of low energy physics to be analogous to the space of genes. On this space there is a "fitness" function, which is the average number of black holes produced by a region of the universe that expands from a bounce. Now, just like the fitness functions of biology, this function is strongly variable, as I said in our universe it is quite large, and there are simple astrophysical arguments that tell us that with many values of the parameters it will be much smaller.

The reason the fitness function is strongly variable is worth mentioning: it is that it is not easy to make a black hole. In our universe, a black hole can only be made if a large amount of matter can be compressed into a very small space, and for this to happen there must be rather special circumstances. The fact that this happens at least once a century in each galaxy of our universe may be said to be due to the fact I described in section II which is that the spiral galaxies are in critical states and so maintain constant rates of star formation over cosmological time scales. Furthermore, the spectrum of masses of stars produced is power law, so that significant numbers of stars are made which are larger than the minimum size by enough of a factor that they can collapse to a black hole even after they supernova and return most of their mass, and sizable amounts of energy to the interstellar medium. For the galaxy to be in the critical state it must be the case, as I mentioned that the rather complicated cooling mechanisms which make possible the giant molecular clouds exist. But this requires that the universe be chemically complex. In short, to a first approximation at least, our universe can overcome the barriers to formation of black holes efficiently because it is chemically complex.

But with this theory we may turn this around and postulate that our universe have the improbable values of the parameters that are necessary for such complexity because this leads to one way to maximize the fitness function, and so make many black holes.

I will not go into details about the statistical arguments, as they are the same as those that are well known to people who work on theories of self-organization. Basically, given the rules as I have introduced them, the probability distribution for the ensemble of universes is peaked around local maximum of the number of black holes. This means that if our universe is typical, it must have parameters that are near a local extrema of the fitness function.

This leads to definite predictions about astrophysics, because it has a simple consequence: all small changes in the parameters from their present values should lead to universes that make less black holes than ours. Thus, the theory is eminently falsifiable; all that would be required to kill it is to find one parameter of the standard models of physics and cosmology, a small variation of which would lead to a large increase in the number of black holes produced. Given that there are on the order of twenty such parameters, and each may be increased or decreased, this gives at least 40 chances to kill this theory.

After several years of trying, I have not found a definite counterexample to this prediction. Unfortunately, with some exceptions [52, 53] every argument for a change of a parameter going one way or another tends to come face to face with some unsolved problem in astronomy. Two examples will suffice to explain this general situation.

Given the fact that the chemistry of "metals" (astronomers call anything heavier than lithium a "metal"), and in particular processes involving carbon and oxygen, seem to play a crucial role in cooling the giant molecular clouds to the point at which massive stars may be formed, it is natural to argue that if the parameters are changed so that such elements are unstable many less massive stars, and hence many less black holes would be made.

The difficulty with this argument is that it is likely that some amount of star formation did take place early, before these elements were created, because there must have been early generations of star formation to get the process started. And at least some of those stars must have been massive enough to supernova, otherwise carbon would never have been found outside of stars. The question is then how many massive stars are made, in the absence of "metals", compared to how many are now. Unfortunately, this is unknown, as all the massive stars made early have by now long been turned into neutron stars or black holes. But it is possible that this question may be answered by future developments in astronomy.

Without metals, star formation may be primarily a fragmentation process [60], that might be modeled fairly simply. It is also not impossible that the power law distribution of masses produced presently by galaxies can be understood in terms of a description of the spiral disks as self-organized critical systems. It is clear in general that the question of the distribution of stellar masses produced, either presently or primordially, is a problem in statistical physics. Of course, if the theory I described here is true, it must be that star formation without metals produces less massive stars then the present processes with metals. It is tempting to make a simple argument that the power law spectra that allow many massive stars to be produced are consequences of self-organized critical phenomena that require a certain chemistry, and hence complexity. But it is also clearly a possibility that such an argument would be too naive.

Let me describe one more prediction made by this theory. One parameter that plays a crucial role in determining the number of black holes produced is the upper mass limit for neutron stars, m_{uml}. A supernova remnant becomes a black hole if it is more massive than this, otherwise it becomes a neutron star. The theory I described must predict that this parameter is as low as possible, consistent with other processes that play a role in star formation. What would be especially interesting is if m_{uml} were under the control of a parameter that played a minimal role in the star formation processes or early universe cosmology, for if this were the case, it could be independently varied and minimized, to maximize the number of supernova remnants that become black holes.

Remarkably, it seems that there is such a parameter: it is the strange quark mass. The reason is that, according to calculations of Brown, Bethe and their collaborators [58], if the mass of the kaon is low enough, the neutron star matter will be dominated by a kaon condensate. This turns out to greatly soften the equation of state from what it would be if the condensate were absent, which in turn lowers m_{uml}. The result is that they predict $m_{uml} = 1.5 M_{solar}$, while conventional equations of state lead to $m_{uml} = 2.5 - 3 M_{solar}$.

If their general arguments are correct, then there is a value of $m_{strange}$, the strange quark mass, $m_{critical}$, such that for $m_{strange} < m_{critical}$ the condensate dominates neutron stars. The question is whether the actual $m_{critical}$ is above the actual value of $m_{strange}$. I may note that the theory I've described here predicts that it must be, for if nature had the possibility of choosing $m_{strange}$ so that many more black holes were produced, and didn't use it, the theory is definitely wrong.

Thus, on this theory I must predict that in fact $m_{uml} = 1.5 M_{solar}$. Thus, the discovery of one neutron star with a larger mass would be strong evidence against it. In fact, of about 18 neutron star masses that are so far measured, all are within error below this value [59].

But there is a second question, why is m_{uml} not still lower? If it were, many neutron stars would instead be black holes. If the theory is true then there must be competing effects that prevent m_{uml} from being lowered still further, even if $m_{strange}$ is lowered. This is a question that can be investigated theoretically, and work on it is underway.

While this discussion has been sketchy, I hope to have convinced the reader that the idea that quantum gravity has no experimental consequences is a bad rap. Here we find that two very plausible assumptions about what happens inside of black holes at the Planck scale result in predictions that can be tested by both observational and theoretical work in astronomy and nuclear physics.

It is quite possible, perhaps even likely that this particular theory is wrong, as I've emphasized if it is wrong we will be able to tell. But we may still learn something from it, for this coupling of assumptions about the Planck scale to predictions about things we can observe is exactly what we may expect if we go away from the idea that the parameters of physics and cosmology are picked by some mathematical principles acting at the Planck scale, and move in the direction of a theory in which they are determined by real mechanisms of self-organization that may have occurred some time in our past.

6 Critical phenomena in quantum gravity and the classical world

Now I would like to come to another way in which critical phenomena are likely to play an important role in cosmology. This application is different from the others I've described, as it involves directly the physics of the Planck scale. As I mentioned earlier, if one assumes that the universe expanded from an initial state, with temperature and densities given by natural units in particle physics, it becomes difficult to understand how the universe managed to expand to the present size, without either collapsing or entering a phase of runaway expansion. However, as I will describe here, the actual situation may be even worse than this. Recent developments in quantum gravity suggest that even the fact that the world has scales in it significantly larger than the Planck scale, which is necessary if it is to be describable in terms of classical geometry, is improbable without fine tunings[13]. Just the fact that there is a world describable in terms of classical space and time, I will argue, is a problem in critical phenomena.

Let me first make the one paragraph argument that this might be the case, then I will show that this argument does in fact correspond to what we know about quantum gravity. A quantum theory of gravity has one scale in it, the Planck scale. Because the scale is also the gravitational coupling constant, what a quantum theory of gravity naturally describes is a strongly coupled phase in which there are no correlations on larger scales. But as a quantum theory of

---

[13] It may be emphasized that in quantum gravity the classical limit is the same as a limit of large distances because $\hbar$ appears only in the Planck length, $l_P = \sqrt{\hbar G_{Newton}}$. Equivalently, it makes no sense to speak of a classical description at the Planck scale.

gravity is a theory of geometry, the existence of a semiclassical limit means that there is a description in terms of a classical geometry in which the averages of classical curvatures are small in Planck units. Thus, classical space and time are themselves consequences of a critical behavior in which there are correlations on scales much larger than the Planck scale. Further, as the coupling of excitations of the geometry are proportional to the wavelength, in Planck units, the existence of a classical limit in a quantum theory of gravity means precisely that the system is critical and weakly coupled. Generically, such a phase cannot exist naturally unless there is some reason for the system to be critical.

Perhaps one might have the impression that this argument proves too much. For what it claims is that in any formulation of quantum gravity in which the existence of classical spacetime is not put in from the beginning, it will be hard to get the classical world out, unless the theory has a critical point for some tuning of the parameters or initial conditions. Formulations of quantum gravity that do not assume that the world is described by small perturbations around a classical spacetime are non- perturbative by definition. And, so far, every non-perturbative formulation that has been developed sufficiently to ask the question leads to the picture I've described.

This has been seen in both path integral and hamiltonian formulations of non-perturbative quantum gravity. In the path integral case, non-perturbative calculations have been performed by discretizing the manifold, and then averaging over a certain set of discrete geometries, as in the case of random surface models in lower dimension [61]. There are two such formulations, the dynamical triangulation models, developed by Agishtein and Migdal [62] and Ambjorn and collaborators [63] that mimic closely the random surface theory and the Regge calculus models [64], which use an older approach in which the dynamical variables are the edge lengths of a fixed triangulations [65].

The results are similar in these two cases. The models have two parameters, which correspond to Newton's constant, G, and the cosmological constant, Λ. There are two phases, a crumpled phase in which macroscopic distances are not defined, and the Haussdorf dimension grows with the size of the system, and an elongated phase, in which things are greatly stretched out, so that the Hausdorff dimension of spacetime is close to 2. Between them there is a second order phase transition governed by a non-trivial critical point at which the Haussdorf dimension seems to be four, within statistical error.

So in these models the picture I described is exactly true. Despite the fact that it is constructed by making a discrete approximation to four dimensional general relativity, the theory can only predicts the existence of a classical four dimensional spacetime when the parameters are tuned to a critical point[14].

A similar picture emerges from the Hamiltonian formulation. Without going into details, one approach to the Hamiltonian quantization of general relativity [67, 68, 69, 70, 71] has advanced to the point that the following simple picture

---

[14] The general idea that the existence of four dimensional quantum gravity would require the presence of a non-trivial scaling behavior associated with a non-Gaussian fixed point was anticipated some time ago on general grounds [66]

has emerged:

The quantum states of the gravitational field are in one to one correspondence with a certain class of graphs, which are called spin networks [72]. These are graphs in which the edges are labeled by half-integers corresponding to spin, and the laws of addition of angular momentum must be satisfied at vertices. It should be emphasized that the graphs are defined only topologically, they are not located anywhere in space, because they are the quantum fields that comprise space.

These states have a simple physical interpretation [71, 73]: they are eigenstates of certain observers that measure the geometry of space by determining the areas of arbitrary surfaces and the volumes of arbitrary regions. Given any such graph, one may draw regions and surfaces and assign them areas and volumes according to simple rules. Every surface has an area given by the sum of the spins on the edges of the graphs that intersect it, in units of the Planck length squared. Every vertex carries a certain discrete amount of volume, given by a certain combinatorial formula of the spins that enter it, times the Planck length cubed.

I want to emphasize that this simple picture was not dreamed up, it is the result of calculations. The fact that the operators that measure physical areas and volumes are discrete is a prediction of quantum general relativity.

Given such a network then, there is a discrete geometry, which is somewhat analogous to those that are integrated over in the path integral approaches (only they correspond to space and not spacetime.) As in that case, almost none of the states of the theory correspond to smooth classical geometries. For certain very special states, based on very large networks which satisfy certain conditions of regularity, it is possible to describe the geometry on the average in terms of a classical metric. But the conditions that make this possible are rather strict, and most of the states of the system do not correspond to any classical geometry, nor do they define any scale of phenomena larger than the Planck scale.

The dynamics under which these networks evolve has been worked out, given certain assumptions about time. This is a long story in itself, let me say only that time here is measured relative to some matter field [74]. The hamiltonian is known, and is a finite, well defined operator [75]. Its action is particularly simple when developed in a strong coupling expansion, in a dimensionless parameter which is $1/G^2 \Lambda$. There are processes that turn vertices into little triangles by adding two new vertices, and processes that do the reverse and collapse little triangles to nodes [76].

The description is very beautiful, and calculations of transition amplitudes can be carried out to any order in this strong coupling expansion. The problem, of course, is that the dynamics in this strong coupling phase does not seem to correspond to the weak coupling picture in which massless gravitons move on a background described by a classical spacetime.

I should emphasize that the problem is not with the existence of gravitons *per se*. It is known, in fact, that if one can assume the existence of a state that has a classical description in terms of a flat geometry, its long wavelength excitations consistent with the gauge invariance and dynamics are precisely two massless spin two modes per momenta [77]. The problem is that the theory does not naturally predict the existence of a state associated with a classical geometry.

I might stress that this is an intrinsically cosmological problem, in that a boundary condition has been imposed in which the universe is spatially compact. This was a condition that Einstein argued for on philosophical grounds, as he invented the science of relativistic cosmology. He was motivated to do so by the philosophical tradition of Leibniz and Mach according to which space and time should not exist a priori, but should be a consequence of dynamical relations among things in the world. What seems to be the case is that when quantum theory is added to the picture this philosophy is realized precisely in that all that one has for generic couplings is a description of a dynamically evolving network of relations. That these have long range correlations such that space and time exist at all has become a dynamical problem, it has become precisely a problem of critical phenomena.

As I said in the introduction, we understand two broad classes of critical phenomena, second order phase transitions and self- organized critical phenomena. The first requires that parameters be tuned to a critical point. But we are discussing a theory that is supposed to be a fundamental theory of cosmology. We might then argue that in such a theory it is not acceptable to explain the existence of the classical world by means of a delicate tuning of parameters. There is nothing outside the world that can tune the parameters. Thus, if it is to succeed, quantum cosmology must become a study of a self-organized critical phenomena. There must be a natural mechanism of self-organization that explains why the quantum state of the world is in an improbable critical state.

Perhaps this may seem too philosophical. But we must keep in my mind that any such theory may be observationally testable, for we may expect generally that if there is a mechanism of self- organization that explains naturally why the world gets big and classical, that mechanism is likely going to produce a scale invariant spectrum of fluctuations around the average state. Thus, such a mechanism is likely to produce an outcome similar to that given by inflationary cosmologies, which is a large classical world on which there is an approximately scale invariant distribution of fluctuations, but, if it succeeds, it will do it naturally, without the fine tunings required to make inflationary models work. As such, it is likely to make testable predictions about the details of the fluctuations seen in the microwave background radiation.

7 Variety, complexity and relativity

It is of course possible that the point of view I've sketched in the last sections will not turn out to be useful. The test of any scientific hypotheses must, in the end, can be nothing other then whether they work out in detail to explain the empirical world. Thanks to the work of the astronomers, cosmology is becoming more and more a question of the details. But, even so, I would like to argue that what is happening deserves some wider reflection. I offer the following as a possible point of view, for whatever it may turn out in the end to be worth.

What we are engaged in is an attempt to make sense of a cosmological theory based on general relativity and quantum theory. This, I would like to argue, must lead to a description of a world that is intrinsically complex, so that the complexity of the world we see must be not accidental, not a matter of a fine tuning of parameters, but in some way inherent in the postulates of quantum theory and relativity.

I know of two arguments for this, one from relativity and one from quantum theory.

The argument that the principles of relativity require a complex world, when applied in a cosmological context is based only on diffeomorphism invariance, which is the most fundamental principle of general relativity. It is the gauge symmetry of the theory, thus it has a more secure status then the particular forms of the dynamical equations. We might expect that it could be included in a larger gauge symmetry in some unification such as the posited non-perturbative string theory, but we cannot expect general relativity to be unified into a more fundamental theory without diffeomorphism invariance.

Diffeomorphism invariance, which Einstein called general covariance, has a very simple meaning in the context of field theory. It says that points have no meaning unless they are described by the values of physical fields. No physical observable can speak about what happens at a point of spacetime, unless that point is determined uniquely by the fields that an observer at that point would measure. You cannot say, what is the curvature scalar at point x. You can only say something like: what the value of the scalar curvature is at a point where the value of the electromagnetic fields (and perhaps their derivatives) are such and such[15].

Like any gauge theory, the physical interpretation of general relativity must be described in terms of gauge invariant observables. As the theory has two degrees of freedom per point, there must be an equal number of such observables. They must all be complicated functions that describe relationships between fields, such as I have described.

Now we come to the key point, which is that such observables will not be well defined for a given cosmological solution to the theory unless it describes a world that is complex enough that points of spacetime can be uniquely described by the values of the fields there. This has a simple consequence, it means that to have a good, gauge invariant interpretation, a spacetime must be complex enough that no two observers observe exactly the same thing, no matter where they are in space and time. To put it more informally, it must be possible to tell where in the world you are, and when it is, uniquely from what you see when

---

[15] There has been in the past some controversy about the question of the interpretation of general relativity, but this view is presently widely understood by relativists to be correct. That it was Einstein's point of view is convincingly shown by Stachel in [78]. This point of view has also been found to be necessary to make progress in quantum gravity [67, 69, 74, 71].

you look around you.

We live in a world with enough variety and structure that this is certainly the case. What I am arguing is that if the gauge invariance of the world includes diffeomorphism invariance this cannot be an accident: it is required if the theory is to have a good interpretation.

There may seem to be a problem with this argument, which is that no solution with symmetry can be given a good physical interpretation by means of such observables, precisely because a symmetry means that there are points that are not distinguished by the values of the fields. But we use solutions with symmetries all the time to model relativistic cosmologies, and we are able to interpret them. Certainly we are, but we do this in a way that makes use of special coordinate systems that are present because of the symmetries. These methods do not generalize to other solutions, nor, I am claiming, can any interpretation that applies generally to relativistic cosmologies be applied to the symmetric solutions.

What we are really doing when we study solutions with symmetries, of course, is taking advantage of the fact that the symmetry is not exact, for it is only by the detailed distribution of matter, that break it, that we are able to give meaning to the coordinates we use.

This circumstance would not be a problem in a Newtonian cosmology, as coordinates are intrinsically meaningful according to the Newtonian conception of space and time. But general relativity is in a different tradition, it is in the tradition of Leibniz and Mach, who argued for a view of space and time in which they are only meaningful to the extent that they are seen in relationships between real things. Indeed, Leibniz understood from the beginning that any cosmological theory in which such a view of space and time was realized would have to describe a world with sufficient complexity that no two observers have exactly the same view of things [79].

The second argument for a complex world, coming from quantum theory has been given by many others, so I will be brief. Quantum theory does not seem to make sense unless there are observers in the world. Therefor, any quantum theory that successfully applies to cosmology must, by self-consistency alone, describe a world complex enough to have observers.

In my opinion, the first argument is stronger than the second. It could easily turn out that quantum theory cannot be extended from the microscopic world to the cosmological. But the first argument uses the most secure principle of general relativity which is diffeomorphism invariance. The observed orbits of the binary pulsars show that we live in a world in which the geometry of spacetime is dynamical, which means there can be no going back to the Newtonian conception of space and time.

However, given either argument we reach the conclusion that a cosmology which is consistent with both general relativity and quantum theory must, by self-consistency alone, describe a complex universe.

If this is, however, to be a good scientific argument, it must be possible to make it quantitative. There ought to be a measure of the complexity of the universe, or of any closed system, that describes how easily each observer may be distinguished by their view of the rest of the system. I would then like to close by describing one such approach to a quantitative measure of complexity, that Julian Barbour and I have been developing.

To define such a notion, we need a system, made of a number of elements, which I will denote x_i. One can think of these as particles or observers, as one likes. What is required is that there be a space $\sqsupseteq$ that contains the possible views of the system. To each element x_i we are able to construct an element, v_i that can be called its view of the system.

For example, the system could be a lattice dynamical system in D dimensions, in which case an element of $\mathcal{V}$ consists of a series of spaces $\mathcal{V}_n$ which describe the possible configurations of neighborhoods of a point in the lattice. n refers to the number of steps away from the original point that describe the neighborhood so that $\mathcal{V}_n$ is the space of possible configurations of a $(2n+1)^D$ lattice of points n steps away from a given site.

Another possibility is that the system is a graph or a network, perhaps of the kind we discussed in the previous section, in which case the neighborhoods $\mathcal{V}_n$ are all the subnetworks with a distinguished point, corresponding to the element, which contain points up to n steps away from it.

Still another possibility is that the system consists of N points distributed in D dimensional space, in which case its view of the rest of the system are $N-1$ points distributed on a $D-1$ dimensional sphere that describes where it sees the other points on its sky.

Given any such system, which defines a set of views w_i of each element, we may define the *variety* of the system as follows.

We must first construct a matrix of differences D_{ij} that measure how far apart the views of the i'th and j'th elements are from each other. There are two approaches to this. The space of views could be a vector space, in which case

$$D_{ij} = |w_i - w_j| \qquad (57)$$

Or, in the cases in which the views comprise a sequence of neighborhoods, the difference D_{ij} is simply $1/n_{ij}$, where n_{ij} is the smallest n such that the two n step neighborhoods are different.

Given the matrix of differences, the variety of the system may be defined.

$$V = \sum_{i \neq j} D_{ij} \qquad (58)$$

We have applied this definition of complexity to a number of systems, including graphs and points in one and two dimensional spaces [80]. We find that systems that have high variety are generically distinguished by being complex without being ordered, so that any two points can indeed by easily distinguished from each other by looking at what is around them. Moreover, this is a definition of complexity that distinguishes true complexity from order, for ordered configurations, or configurations with any kind of symmetry turn out to have low variety. Generically, we find that ordered configurations have much lower varieties than randomly generated configurations, while configurations with high variety are easily distinguished from both ordered and random configurations.

Thus, the variety of a system may be defined quantitatively. The next step is to try to define an appropriate notion of variety for classical or quantum general relativity. We may, for example, try to define the variety of a quantum spacetime to be inversely proportional to the average number of bits of information an observer must have in order to locate themselves uniquely in space and time. We may then conjecture that the dynamics of a quantum gravitational theory act to increase the variety of typical configurations in time. Certainly, as gravitation acts to form hierarchies of bound systems, as we see from the Press-Schecter model, and more generally makes it possible for large regions of the world to be kept far from thermal equilibrium for arbitrarily long periods, this is not obviously wrong. If true, this would be a step towards a picture in which we understood that our world is organized because a quantum gravitational system must, for its own self-consistency, contains intrinsic statistical mechanisms of self-organization.

ACKNOWLEDGEMENTS

I want to thank the organizers of the Guanajuato school for giving me the opportunity to describe these thoughts in front of a knowledgeable and critical audience. I want to also thank the audience and lecturers for many illuminating discussions and criticisms. Some of the ideas described here have been developed over a long period of time through conversations and collaborations with Julian Barbour, Jane Charlton, Louis Crane and Carlo Rovelli. I would also like to thank also Per Bak, Gerry Brown, Saint Clair Cemin, Freeman Dyson, James Peebles, Martin Rees, Stanley Rosen, Larry Schulman and John Wheeler for helpful discussions about these problems. I am also grateful to Jane Charlton for a critical reading of the manuscript. Finally, I would like to thank the astrophysicists of the Institute for Advanced Study, where these notes were written, for an atmosphere most conducive to reflection on the unsolved problems in cosmology. This work has been supported in part by the NSF grant PHY-93-96246.

References

1. S. D. M. White, J. F. Navarro, A. E. Evrard, C. S. Frenk, Nature 366 (1993) 429-433;
2. P. Coles and G. Ellis *The case for an open universe* Department of Applied Math prprint, Capetown (1994).
3. See for example, J. Bachall, in the Proceedings of Some Unsolved Problems in Astrophysics.
4. Mather, J. C. et al. Ap. J. 354 (1990) L37.
5. V. De Lapparent, M.J. Geller and J.P. Huchra Ap.J. 302 (1986) L1; M. P. Haynes and R. Giovanelli, Ap. J. 306 (1986) L55.

6. L J For an excellent review, see Garay: "Quantum gravity and minimum length", Imperial College preprint/TP/93-94/20, gr-qc/9403008 (1994)

7. F. Hoyle, D. N. F. Dunbar, W. A. Wensel and W. Whaling, Phys. Rev. 92 (1953) 649; F. Hoyl, *Galaxies, Nuclei and quasars* (Heinemann, London, 1965), p. 146.

8. B. J. Carr and M. J. Rees, Nature 278 (1979) 605.

9. J. D. Barrow and F. J. Tipler *The Anthropic Cosmological Principle* (Oxford University Press,Oxford,1986).

10. B. Carter, *"The significance of numerical coincidences in nature"*, unpublished preprint, Cambridge University, 1967; in *Confrontation of Cosmological Theories with Observational Data*, IAU Symposium No. 63, ed. M. Longair (Reidel, Dordrecht,1974) p. 291.

11. P. Bak, C. Tang and K. Wicsenfeld, Phys. Rev. A38 (1988) 364; Phys. Rev. Lett. 59 (1987) 381.

12. P. Bak and Maya Paczuski, "Complexity, contingency and criticality" Brookhaven preprint.

13. S. Mineshinge, N. B. Ouchi and H. Nishimori, PASJ 46 (1994) 97; S. Mineshinge, M. Takeuchi and H. Nishimori, Ap. J. 435 (1994) L125.

14. H. Gerola and P. E. Seiden, Ap. J. 223 (1978) 129; P. E. Seiden, L. S. Schulman and H. Gerola, *Stochastic star formation and the evolution of galaxies*, Astrophys. J. 232 (1979) 702-706; P. E. Seiden and L. S. Schulman, *Percolation and galaxies* Science 233 (1986) 425-431 *Percolation model of galactic structure*, Advances in Physics, 39 (1990) 1-54; L. S. Schulman, "Modeling galaxies:: cellular automana and percolation", to appear in *Cellular Automata: Prospects in Astrophysical Applications*, A. Lejeune and J. Perdang, eds. World Scientific, Singapore (1993).

15. P. Seiden, L.S. Schulman and H. Gerola, Ap. J. 232 (1979) 702.

16. See, for example: J. Franco and D. P. Cox, *Self-regulated star formation in the galaxy* Astrophys. J. 273 (1983) 243-248; J. Franco and S. N. Shore *The galaxy as a self-regulated star forming system: The case of the OB associations* Astrophys. J. 285 (1984) 813-817; S. Ikeuchi, A. Habe and Y. D. Tanaka *The interstellar medium regulated by supernova remnants and bursts of star formation* MNRAS 207 (1984) 909-927; R.F.G. Wyse and J. Silk *Evidence for supernova regulation of metal inrichment in disk galaxies* Astrophys. J. 296 (1985) 11-15; M. A. Dopita, *A law of star formation in disk galaxies: Evidence for self- regulating feedback* Astrophys. J. 295 (1985) L5-L8; G. Hensler and A. Burkert, *Self-regulated star formation and evolution of the interstellar medium* Astrophys. and Space Sciences 171 (1990) 149- 156.

17. R. F. G. Wyse and J. Silk, Astrophys. J. 339 (1989) 700.

18. See, for example, *The Physics and Chemistry of Interstellar Molecular Clouds* ed. G. Winnewisser and J.T. Armstrong, Springer Verlag Lecture Notes in Physics 331(1989); *Molecular Coulds in the Milky Way and External Galaxies* ed. R. L. Dickman, R. L. Snell and J. S. Young Springer Verlag Lecture Notes in Physics 315 (1988).

19. B. G. Elmegreen *Triggered Star Formation* IBM Research Report, in the Proceedigs of the III Canary Islands Winter School, 1991, eds. G. Tenorio-Tagle, M. Prieto and F. Sanchez (Cambridge University Press, Cambridge, 1992).

20. B. G. Elmegreen, *Large Scale Dynamics of the Interstellar Medium*, to appear in *Interstellar Medium, processes in the galactic diffuse matter* ed. D. Pfenniger and P. Bartholdi, Springer Verlag, 1992.

21. A. Parravano, *Self-regulating star formation in isolated galaxies: thermal instabilities in the interstellar medium* Astron. Astrophys. 205 (1988) 71-76; *A self-*

regulated star formation rate as a function of global galactic parameters Astrophys. J. 347 (1989) 812-816; A. Parravano and J. Mantilla Ch., *A self-regulated state for the interstellar medium: radial dependence in the galactic plane*, Atrophys. J. 250 (1991) 70-83; A. Parravano, P. Rosenzweig and M. Teran, *Galactic evolution with self-regulated star formation: stability of a simple one-zone model* Astrophys. J. 356 (1990) 100-109.

22. See, for example, F. H. Shu, F. C. Adams and S. Lizano *Star formation in molecular clouds: observation and theory* in Ann. Rev. Astron. Astrophy. 25 (1987) 23-81 and C. J. Lada and F. H. Shu, *The formation of sunlike stars* Berkely preprint, to appear in Science and references contained therein.

23. E. E. Salpeter, Astrophys. J. 121 (1955) 161.

24. G. E. Miller and J. Scalo, Ap. J. Suppl. 41 (1979) 513; J. Scalo, Fundamentals of Cosmic Physics 11 (1986) 1-278.

25. R. B. Larson M.N.R.A.S. 214 (1985) 379; 218 (1986) 409.

26. J. Scalo, in *Physical Processes in Fragmentation and Star Formation* ed. R. Capuzzo-Dolcetta, C.Ciosi and A. Di Fazio (Klower,1990)

27. B. Elmegreen and M. Thomasson, *Grand design and flocculent spiral structure in computer simulations with star formation and gas heating* Astron. and Astrophys. (1992) ?.

28. M. Bucher, A. S. Goldhaber and N. Turok, "An open universe from inflation", hep-ph/9411206, iassns-hep-94-81. PUPT-94-1507; M. Bucher and N. Turok, "Open inflation with arbitrary false vacuum mass" hep-ph 9503393, PUPT-95-1518; J.R. Gott, Nature 295 (1982) 304.

29. Walker, T. P, Steigman, G., Kang, H.-S., Schramm, D. M., & Olive, K. 1991,

30. See, for example, S. D.M. White, MAP preprint, 1994; S. D. M. White and C. S. Frenk, Ap. J. (1991) 379, 52; G. Efstathios and J. Silk, Fund. Cos. Phys. 9 (1983) 1.

31. J.P. Ostriker and L.L. Cowie, Ap.J. 243 (1981) L127.

32. S. Ikeuchi, Publ. Astron. Soc. Japan 33 (1981) 211.

33. J.P. Ostriker, C. Thompson and E. Witten, Phys. Lett. B (1986).

34. R.A. Daley (1986)

35. N. Yu. Gnedin and J. P. Ostriker, Astrophys. J. 400 (1992) 1-20

36. P.J.E. Peebles, in *The Early Universe* ed. W.G. Unruh and G.W. Semenoff, D. Reidal Publishing, 1988, p. 203; in the proceedings of the 8th IAP meeting, *First light in the universe*

37. C. Hogan, Ap. J. 415 (1993) L63-66.

38. P. Petitjean, J. K. Webb, M. Rauch, R.F. Carswell and K. Lanzetta, MNRAS 262 (1993) 499; K.M. Lanzetta, A. M. Wolfe, D. A¿ Turnshek, Limin Lu, R.G. McMahon and C. Hazard, Ap. J Suppl. Series. 77 (1991) 1.

39. J. Charlton, E. Salpeter and C. J. Hogan, Ap. J. 402 (1993) 493; J. Charlton, E. Salpeter and S. M. Linder, "Competition between pressure and gravity confinement in Lyman alpha forest observations", ApJ, 430, L29 (1994).

40. Lanzetta, K. M., Bowen, D. V., Tytler, D., and Webb, J. K. 1995, ApJ, 442, 538; Steidel, C. 1995, in Proceedings of ESO Workshop on QSO Absorption Lines, ed. G. Meylan, (Springer-Verlag: Heidelberg), in press

41. Richard Ellis "The morphological evolution of galaxies" in *Unsolved Problems in Astrophysics op. cit.*

42. Broadhurst, R. J., Ellis, R. S., & Shanks, T. 1988, MNRAS, 235, 927 Colless, M. M., Ellis, R. S., Taylor, K., & Hook, R. N. 1989, MNRAS, 244,408 Songaila, A., Cowie, L. L., Hu, E. M., & Gardner, J. P. 1994, ApJS, 44, 461

43. J. Peebles, *Principles of Physical Cosmology* (Princeton University Press,1993).

44. For a general review, see N.Bachall and J. Ostriker, in the Proceedings of the Conference on Unsolved Problems in Astrophysics, *op. cit.*. See also R.Y. Cen and J. P. Ostriker, Ap. J. 339 (1992) L113; 404 (1993) 415; 417 (1993) 415; D. Ryu, J.P. Ostriker, H. Kang and R.Y. Cen, Ap.J. 414 (1993) 1

45. M. Carfora and K. Piotrokowska, "A renormalization group approach to relativistic cosmology", to appear in Phys. Rev. D.

46. W. H. Press and P. Schecter, Ap. J.. 187 (1974) 425.

47. S. D. M. White, G. Efstathiou and C. S. Frenk, Mon. Not. R. Astro. Soc. 262 (1993) 1023; A. Klypin, J. Holtzmann, J. Primack and E. Regos Ap. J. 416 (1993) 1; Lacey and Cole, MNRAS (1994)

48. K. Chen and P. Bak, Phys. Lett. A 140 (1989) 299.

49. L. S. Schulman and P. E. Seiden, Ap. J. 311 (1986) 1.

50. B.J. Carr, "Baryonic dark matter", to appear in *Annual Reviews of Astronomy and Astrophysics*, 1995.

51. A. Strominger, "Massless black holes and conifolds in string theory", preprint hep-th/9504090.

52. L. Smolin Classical and Quantum Gravity 9 (1992) 173-191

53. L. Smolin, *On the fate of black hole singularities and the parameters of the standard model* gr-qc ??

54. L. Smolin, *The Life of the Cosmos* to appear in Oct. 95, Crown Press,New York, and Orion Press, London.

55. J.A. Wheeler, in *Gravitation*, by C. Misner, K. Thorne and J. A Wheeler, last chapter.

56. E. Martinec, 1994, hep-th/9412074

57. R. Gott, private communication.

58. G. E. Brown and H. A. Bethe, Astro. J. 423 (1994) 659; 436 (1994) 843, G. E. Brown, Nucl. Phys. A574 (1994) 217; G. E. Brown, "Kaon condensation in dense matter"; H. A. Bethe and G. E. Brown, "Observational constraints on the maximum neutron star mass", preprints.

59. S. E. Thorsett, Z. Arzoumanian, M.M. McKinnon and J. H. Taylor Astrophys. Journal Letters 405 (1993) L29

60. M. Rees, MNRAS 176 (1976) 483; J. Silk Ap.J. 211 (1976) 638.

61. For a review, see J. Ambjorn, J. Jerkiewicz and Y. Watabiki, "Dynamical triangulations, a gateway to quantum gravity" NBI-HE-95-08, to appear in J. Math. Phys. Nov. 1995.

62. M.E. Agishtein and A. A. Migdal, Nucl. Phys. B385 (1982) 395.

63. J. Ambjorn, J. Jerkiewicz and C. F. Kristjansen, Nucl. Phys. B393 (1993) 601; Phys. Lett. B305 (1993) 208; J. Ambjorn, Z. Burda, J. Jerkiewicz and C. F. Kristjansen, Phys. Rev. d48 (1993) 3695.

64. H. W. Hamber, Nucl Phys. B (Proc. Supp.) 20 (1991) 728; 25A (1992) 150; B400 (1993) 347; Phys. Rev. D45 (1992) 507; H. W. Hamber and R. M. williams, Nucl. Phys. B415 (1994) 463.

65. T. Regge, Nuovo Cimento 19 (1961) 558.

66. S. Weinberg, in *General Relativity: An Einstein Survey* ed. S. Hawking and W. Israel (Cambridge University Press,1979). L. Smolin, Nuclear Physics B208 (1982) 439.

67. C Rovelli: Class Quant Grav 8 (1991) 1613

68. A Ashtekar: *Non perturbative canonical gravity*, World scientific, Singapore 1991

69. L Smolin: in *Quantum Gravity and Cosmology*, eds J Pérez-Mercader *et al*, World Scientific, Singapore 1992

70. A Ashtekar C Rovelli L Smolin: Phys Rev Lett 69 (1992) 237

71. C. Rovelli and L. Smolin, Discreteness of volume and area in quantum gravity, to appear in Nucl. Phys. B 1995.

72. R Penrose: in *Quantum theory and beyond* ed T Bastin, Cambridge U Press 1971; in *Advances in Twistor Theory*, ed. L. P. Hughston and R. S. Ward, (Pitman,1979) p. 301; in *Combinatorial Mathematics and its Application* (ed. D. J. A. Welsh) (Academic Press,1971).

73. C. Rovelli and L. Smolin, "Spin networks and quantum gravity" Penn State CGPG-95/4-4 and IASSNS-HEP-95/27 preprint, gr-qc/9505006.

74. L Smolin: in *Directions in General Relativity, v. 2, papers in honour of Dieter Brill*, ed BL Hu T Jacobson, Cambridge University Press, Cambridge 1994

75. C. Rovelli and L. Smolin, Phys Rev Lett 72 (1994) 446

76. R. Borissov, C. Rovelli and L. Smolin, *Nonperturbative dynamics of quantum general relativity* preprint in preparation. C. Rovelli, to appear in J. Math Phys. Nov. (1995).

77. J Iwasaki C Rovelli: Int J of Mod Phys D1 (1993) 533; Class and Quantum Grav 11 (1994) 1653

78. J. Stachel, *Einstein's search for general covariance 1912-1915* in *Einstein and the History of General Relavity* ed. by D. Howard and J. Stachel, Einstein Studies, Volume 1 (Birkhauser,Boston,1989).

79. Leibniz, *The Monadology* in *Leibniz, Philosophical Writings* ed. G.H.R. Parkinson, translated by M. Morris and G.H.R. Parkinson (Dent,London,1973)

80. J. B. Barbour and L. Smolin, Syracuse University preprint, SU-GP-92/2-4, see also J. B. Barbour, "On the origin of structure in the universe," presented at the "3d Philosophy and Physics Workshop", Forschungsstatte der Evangelischen Studiengemeinschaft (FEST) in Heidelberg, May 1990. To be published in "Philosophy and Modern Physics", publ. by Springer; *Mathematical Modeling of the Monodology*, submitted for publication and L. Smolin *Space and Time in the Quantum Universe* in *Conceptual Problems of Quantum Gravity* ed. by A. Ashtekar and J. Stachel, (Birkhauser,Boston,1991).

Springer-Verlag
and the Environment

We at Springer-Verlag firmly believe that an international science publisher has a special obligation to the environment, and our corporate policies consistently reflect this conviction.

We also expect our business partners – paper mills, printers, packaging manufacturers, etc. – to commit themselves to using environmentally friendly materials and production processes.

The paper in this book is made from low- or no-chlorine pulp and is acid free, in conformance with international standards for paper permanency.

Lecture Notes in Physics

For information about Vols. 1–425
please contact your bookseller or Springer-Verlag

New Series m: Monographs